SCIENCE AND ENGINEERING POLICY SERIES

Fisheries Resources of the Sea and their Management

DAVID CUSHING

OXFORD UNIVERSITY PRESS 1975

Oxford University Press, *Ely House, London W.1*

Glasgow
New York
Toronto
Melbourne
Wellington
Cape Town
Salisbury
Ibadan
Nairobi
Dar es Salaam
Lusaka
Addis Ababa
Bombay
Calcutta
Madras
Karachi
Lahore
Decca
Kuala Lumpur
Singapore
Hong Kong
Tokyo

ISBN 0 19 858320 6

Reproduced and printed by photolithography and bound in Great Britain at The Pitman Press, Bath

Contents

1 The development of fisheries

1. Introduction

Men have eaten fish since earliest times and for centuries they have gone to sea to catch them. Indeed, line fishermen sailed to Iceland in summertime from Dunwich on the Suffolk coast of England as long ago as the fifteenth century; and during the same period, herring fishermen worked from Holland on grounds all over the North Sea. However, fishing did not expand quickly until industrialization was well under way. When the railway reached Grimsby in the middle of the nineteenth century, sailing smacks moved there from the Thames. The fish was brought on ice from sea and was carried on ice by rail to the expanding towns of the English Midlands. As the nineteenth century progressed, the smacks extended their exploration of the North Sea with beam trawls from the Chops of the Channel to the Northern North Sea. In 1881, the steam trawler with its double-barrelled steam winch was invented, and by the turn of the century the sailing smacks were more or less replaced, if not completely so. By then the density of fish had fallen by a third. The consequent loss of profit to the fishermen led eventually to the international controls that will be described in later chapters.

There was a similar history of industrialization of the fisheries in North America. Between 1870 and 1880, the mackerel and menhaden boats off the eastern seaboard of the United States were fitted with steam engines. The oil extracted from the fish was used for saddle soap in the expansion of the West. In the first decade of the present century, somewhat similar vessels exploited the Pacific halibut with hook and line from Seattle and Vancouver. The development of this fishery depended upon the establishment of cold stores when the transamerican railways were opened. The North Sea history was repeated as stock densities and profits declined, to be followed by the creation of an international commission for the control of the fishery, as will be described below.

The Victorian pace of industrialization was relatively slow as compared with contemporary rates of change. It is no accident that the events in microcosm at the end of the last century and the beginning of this one are now being repeated in macrocosm. The successful fleets that spread across the North Sea and along restricted parts of the North American coastline have now extended their activities to all oceans. Indeed the major problem facing us today is not one of exploration but one of conservation, because most of the world's fish stocks are now being exploited.

2. The ways in which fish are caught

There are four main methods of catching fish: first, by hook and line; secondly by gill net or drift net; thirdly by an encircling net (purse seine, ring net, or lampara); and lastly by trawl. The hook and line is the traditional gear familiar to everyone; the Great Long Lines used by British fishermen in northern waters had baited hooks spaced perhaps six feet apart along two to three miles of line laid upon the sea bed. In the subtropical tuna fishery, unbaited hooks are suspended from a buoyed line up to 80 km long at the surface. The gill net or driftnet is a curtain of net hung from floats at the surface; herring or salmon swim into it and are caught by their gill covers. The drift net used by the East Anglian herring fishermen was seven fathoms deep and thirty-five fathoms long and the nets were linked in 'fleets' of up to two and a half miles in length by a messenger rope. The rope was attached to the fishing vessel, or drifter, and the whole system drove with the tide. Such nets were used to catch herring in the North Sea until the late nineteen-sixties and are still used to catch Asiatic salmon in the open Pacific. Sometimes gill nets were laid on the sea bed to float upwards and catch herring or cod close to the bottom.

There are many varieties of encircling net by which a shoal of fish is surrounded and caught. A purse seiner locates such shoals fairly near the surface and the net is closed with pursing rings below them so that they cannot escape downwards. When the fish are retained within the ring they are lifted from the water with brailers or large dip nets and are then carried aboard the purse seiner. Ring nets and lamparas are somewhat smaller and are used to catch fish quickly without pursing, i.e. the ring is closed at the surface. Up till the end of the last century, pilchards were caught on the coasts of Cornwall in south-western England by very large shore seines which employed the principle of the encircling net on the beaches.

A trawl is a conical bag of net dragged across the sea bed to catch the fish on it or close to it. The trawls towed by sailing smacks in the last century were beam trawls, which were kept open by long ash beams between the iron trawl heads or runners. Later, when the smack was superseded by the steam trawler, the beam trawl was replaced by the otter trawl. The wings of the otter trawl are spread apart by the otter boards, or doors (which work like kites) and by the 'dan lenos' (or *guindineaux,* or *guiders*). Trawls are the major instruments of capture on the sea bed throughout the world ocean. In recent years, Dutchmen have returned to the use of beam trawls for catching sole and plaice in the southern North Sea because they dig into the sand a little and because as one is hauled on one side of the ship, a second can be shot on the other side. A development of the trawl, the Danish seine, is used as an encircling net on the bottom; long coils of rope are laid from the wings of the net in a large circle, which is closed as the ropes are hauled and the fish are driven into the net. The gear is light and can be handled by three or four men in a rather small boat, whereas a trawl tends to be heavier and has to be handled by a larger crew in a bigger ship.

Today, the four main types of gear are highly developed. The Japanese and Korean pelagic long-liners catch the tuna-like fishes in the tropical and subtropical waters of all the world oceans. The fishermen spend long periods at sea and land their catches into cold stores in various parts of the world to supply the avid market in the United States for frozen tuna. Large purse seines have been used by Norwegian and Icelandic fishermen to catch herring as much as 150 m from the surface in the late autumn in the East Icelandic Current. All such purse seiners are equipped with echosounders and sonars for locating fish shoals in midwater. The large stern trawlers, which are often freezers also, that work on the sea bed in all oceans are armed with large trawls of considerable spread. All such gears, pelagic long lines, the long surface gill nets, big and deep purse seines, and large trawls have considerable capacity to kill fish, which is at the same time a capacity to make large and profitable catches.

The ships today are larger than they used to be; they can stay at sea for longer and can cross from one side of an ocean to the other. The trawlers are more powerful, can tow bigger trawls, and can work in worse weather. The purse seiners of twenty years ago worked fairly close to the shore because the net was shot from a dory; today, it is shot from the seiner itself and is hauled with a large power block. As a consequence, such vessels can work in the open sea in poor weather, indeed in the open Atlantic during the winter-time. The tuna purse seiners from

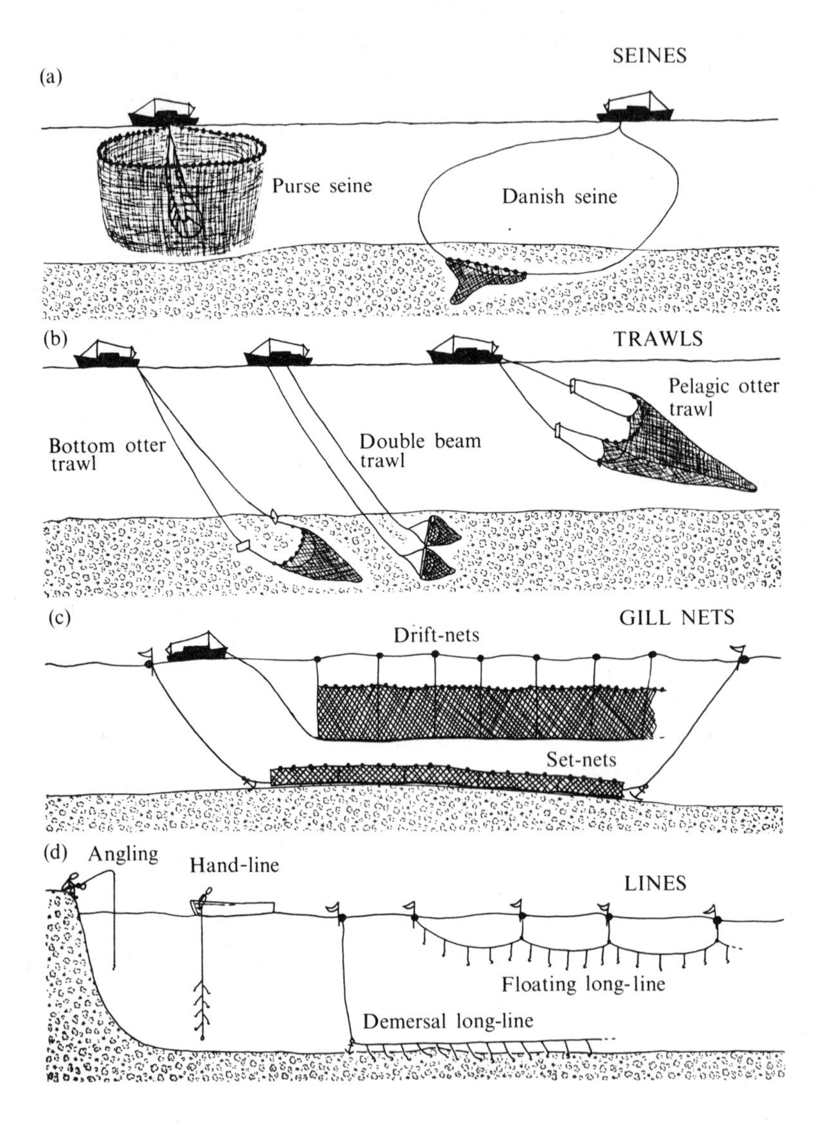
SEINES
(a)
Purse seine
Danish seine
(b)
TRAWLS
Pelagic otter trawl
Bottom otter trawl
Double beam trawl
(c)
GILL NETS
Drift-nets
Set-nets
(d) Angling
Hand-line
LINES
Floating long-line
Demersal long-line

northern Norway. Other stocks exploited for fish meal included the menhaden off the eastern seaboard of the United States, the British Columbian herring, the pilchards off South and South West Africa, and the anchoveta off Peru. The fishery for the anchoveta is the largest in the world and has yielded about ten million tons each year for a number of years until 1972.

There is a variety of markets of less importance than the four major ones. For example much of the yield of the Californian sardine was canned during the thirties and forties as described so vividly by Steinbeck in his novel *Cannery Row*. Much of the output of the South West African pilchard was canned in the early stages and today the Russians put sprats in cans at sea in the North Sea. For the herring fisheries in the north-east Atlantic there were many outlets, for example, smoked fish such as kippers and red herring, marinated herring and, in Holland, cured herring; they still exist but their importance has declined greatly and some of the ancillary trades have disappeared. Smoked fish remains a valuable market in various specialist trades, for example, smoked eel, smoked salmon, and finnan haddock in European markets.

The distribution of markets throughout the world is of considerable interest. In the North Atlantic trawled fish, frozen or preserved on ice, are of predominant importance—primarily cod, haddock, and some red-fish, although in recent years fish such as silver hake have been caught in the north-west Atlantic. Herring are still caught for fish meal in the west, but with the extinction of most of the herring fisheries in the north-east Atlantic the fish meal market is supplied by capelin and to some degree by mackerel. In the North Pacific, the largest catches are of Alaskan pollack, most of which is converted to fish meal; king crab and salmon from the same areas are frozen. Hake are caught by Russian trawlers off the west coast of the U.S. and a variety of species are taken in the western Pacific. Japanese anchovy off Hokkaido provide a source of fish meal, as do the pollack in the Bering Sea.

The most spectacular fisheries are those for fish meal on herring-like or sardine-like fishes in the upwelling areas, which lie in the subtropical eastern boundary currents. The Californian sardine fishery and the potential one for Californian anchovy are such examples from the Californian current. Off north-west Africa in the Canary current, quite large catches of *Sardinella* are obtained. The Peruvian anchoveta fishery occurs in the Peru current and the sardine and anchovy stocks off southern Africa are exploited in the Benguela current. The total yield from such fisheries amounts to about 15 million tons per year, all of

Table 1

The catches of fishes of different groups of oceanic areas (thousands of tons)

	Flounders, halibuts, soles	Cods, hakes, haddocks	Redfishes, basses, congers	Jack mackerels, mullets, sauries	Herrings, sardines, anchovies	Tunas, bonitos, billfish	Mackerels, snoeks, cutlassfish	sharks, rays ratfish
North-west Atlantic	298	1524	353	14	859	18	398	39
North-east Atlantic	366	3964	691	295	1889	53	378	103
West central Atlantic			201	47	635	46	2	8
East central Atlantic	13	49	294	549	957	209	300	17
Mediterranean and Black Sea	11	32	102	81	431	37	13	14
South-west Atlantic	7	134	177	68	184	12	9	24
South-east Atlantic	3	1121	88	375	1102	22	72	6
West Indian Ocean	10	7	324	69	331	136	149	116
East Indian Ocean	1		89	26	58	49	33	29
North-west Pacific	310	78	783	666	961	336	1558	54
North-east Pacific	257	1623	307		114	20		3
West central Pacific	6		265	422	158	186	198	30
East central Pacific	6		29	27	222	317	1	8
South-west Pacific	1	37	49	31		82	19	5
South-east Pacific	1	99	25	198	5638	71	9	10
Total	1228	11472	3774	2870	13535	1594	3131	466

which is converted to fish meal in addition to those sources of meal in the North Atlantic and North Pacific.

Table 2

The distribution of outlets in the world ocean in 1972

Market	Millions of tons per year
Fresh or preserved on ice	10·5
Frozen	10·8
Smoked, sun-dried, or salted	8·1
Canned	6·8
Fish meal	19·4

The present world catch amounts to about 65–70 million tons per year (the figure is quoted imprecisely because it is not known to what extent the Peruvian anchoveta stock will recover from its recent collapse). Table 1 shows the catches of different groups of fish in different oceanic areas in thousands of tons for the year 1972. The important groups are the cod-like fish and the herring-like ones and the North Atlantic and North Pacific are the areas of most intense exploitation; exceptions are hake and sardine in the south-east Atlantic (in the Benguela current), anchoveta in the south-east Pacific (in the Peru current), and fairly high catches of sardines or sardinellas in the central Atlantic. The group of redfishes are nowhere very abundant but they are spread fairly evenly at all latitudes and in most oceans. Tunas live in subtropical and tropical waters, with mackerels in rather higher latitudes, like the flatfish. Table 2 gives the distribution of markets in millions of tons per year in the world ocean.

The older markets for smoked, sun-dried, and salted fish remain important in tropical and subtropical areas, but in the higher latitudes a large proportion is sold fresh, frozen, or preserved on ice. From the upwelling areas, catches are predominantly converted to fish meal and they remain the major source of fish meal on the world ocean.

4. The need for fish

The need for fish arises from the need for protein in man's diet. This is required both for maintenance and growth and is supplied by vegetable protein and by animal protein in the form of fish and meat. Vegetable protein lacks two amino acids and traditionally fish has been cheaper than meat. In pre-industrial societies, fish was often readily caught rather cheaply whereas meat was reared at some cost; a snobbish judgement in

Britain used to suggest that herring (like rabbits) were food fit only for the poor. Today, the demand for fish ranges from the physical need for protein to the luxury requirements of city restaurants.

The proportion of fish to meat in Britain today is about 10 per cent, having declined from 15–20 per cent after the first world war. The quantity in calories per head of fish protein throughout the countries of the world shows a dome shaped curve as a function of gross national product. In other words, the poorest countries rely on vegetable protein with a minimal quantity of fish and the richer countries depend much more on meat; indeed there may be a tendency in such parts of the world to use fish only as a luxury, in restaurants, rather than as fish fingers or as fish and chips. The consumption of fish appears to be greatest in a middle range of countries in terms of gross national product; that is, those in an early stage of industrialization. Indeed, the expansion of fisheries in Europe increased with the elaboration of the railway network and it would have been quicker to supply the increasing urban population with fish than to intensify stock rearing on the land. There is a sense in which fish and chips was the first convenience food of the industrial age, dependent on railways and plenty of ice and very useful to the hard-worked urban housewives.

In the early twenties, the Japanese decided to expand their fisheries because their expanding population could not be fed by meat grown on the islands. In the same period a similar decision was made in the U.S.S.R. After considerable exploratory voyages, the commercial fleets were expanded in both countries. Today they are the most important countries catching fish, each landing about seven million tons in each year. Much of their catches are taken from grounds distant, or very distant, from the home ports. Perhaps this deliberate expansion of two countries in the midst of extensive industrialization can be properly compared with that when industry expanded in Britain and other European countries towards the end of the last century.

2

The natural history of fishes

1. Age, growth, and fecundity of fishes

For their comparatively small size, fish live for rather a long time. A life span of 10, 20, or 30 years is common among commercial species such as cod, herring, or plaice. In temperate water species, annual rings on otoliths (or earstones) and on scales can be distinguished quite easily, but not so readily in species from tropical seas. The extremes in age determined in this way are considerable; for example an otolith in a sturgeon from Lake Winnebago in Wisconsin had 152 annual rings, whereas the Australian whitebait has none because it lives for only 9 months. Most of the evidence on the lifespan of fishes comes from exploited stocks and therefore represents an underestimate of the ages that might be reached in unexploited ones.

Herring in the North Sea in the unexploited state probably live for 10–15 years and those in the Norwegian Sea, which are somewhat larger, probably live for 20–25 years. The Pacific salmon, of which there are five species, live for 2 or 4 years with a few at 3, 5, or 6 years of age; they die, after spawning for the only time, on their native redds. Sprats live for about 3 or 4 years and plaice up to 35 years. The ages of tuna are unknown but it is not likely that they live for much more than a decade. If fish survive to old age they may well become senescent. After a given size, that of late middle age, the myofibrils of the white muscles decrease markedly in size; as the white muscles are those responsible for the acceleration in escape or attack, old fish must be more vulnerable to predators and less able themselves to find food. There is a little evidence that plaice in the southern North Sea might suffer a senescent mortality and some flatfish off Newfoundland, at the start of the fishing there, were called the watery plaice perhaps because the white muscles had atrophied. To summarize, the lifespan of fishes is quite long and some may die of old age.

Many fish eggs are small, about 1 mm across (but those of the halibut and salmon are 3–4 mm across) and they are usually laid in midwater; herring, sandeels, salmon, and lumpsuckers lay their eggs on the bottom. The salmon lay theirs on gravel in lakes or streams. In marine fishes the yolk lasts for a period of between 48 hours to about three weeks, depending on temperature. Fecundity is a function of weight and the full gonad weighs about one sixth of the body weight. The number of eggs per female ranges between 10^3 and 10^7. As the eggs are laid they absorb water up to five times their volume.

Fish grow quickly in juvenile life but more slowly with maturation and much more slowly in old age. But they grow continuously and during the period of mature life they may grow by a factor of many times, in some cases by an order of magnitude or more. Any study of the effect of fishing concerns itself largely with the maximization of gains in weight in face of the loss of numbers during the life history in the fisheries; if fishing starts on immature fish, the increment of growth during the fishable lifespan is much greater. Sprats grow to about 20g, herring to 100 or 300g, depending on where they live, plaice to about 2 or 3 kg, cod to about 10 kg and some tuna to 100 kg or more. The total increment in growth from egg to adult ranges between 10^4 and 10^8, which is of the same order as the range in fecundity. This is not surprising, because, of the eggs produced by one gonad all but two die when the survivors are somewhere near their greatest size. Two gonads produce two adults and the total increment from generation to generation is a factor of six (the ratio of body weight to gonad weight).

Some fish grow slowly and others quickly. For example, cod off Labrador grow to a length of 70 cm in about nine years whereas in the Celtic Sea cod of another stock reach the same size in four years. The weight of a large cod of 70 cm in length and upwards ranges between one and ten kg. In contrast, the bluefin tuna, a subtropical animal, puts on 50 kg in the first three years of life. In warmer water, animals grow faster than in colder water and the growth rate tends to vary inversely with the final weight; in other words, at the same water temperature bigger fish put on weight more slowly than little ones. However, as a somewhat imperfect rule, fish in the cooler higher latitudes grow more slowly to greater sizes than in the warmer and lower ones. Similarly they tend to live longer.

At the end of the mature life of a larger cod, the two parent gonads are replaced as in other animals, but in such a fecund fish the death rate of eggs, larvae, immatures, and of adults is 99·99998 per cent.

2. The migration circuits of fishes

Most commercial species of fish migrate over considerable distances, up to hundreds or thousands of miles a year. The adults travel between spawning and feeding ground and back again in a regular current. The larvae drift away from the spawning ground to a nursery ground and this passive migration has been named the 'larval drift'; on the nursery they are more or less secure from predation for a year or so while they put on weight. As they mature they move towards the feeding ground where they recruit to the adult stock. A convenient way of representing the migration circuit is in the form of a triangle with spawning ground and nursery ground forming the base and the feeding ground at the apex (Fig. 2). The circuit from spawning ground to nursery ground and thence

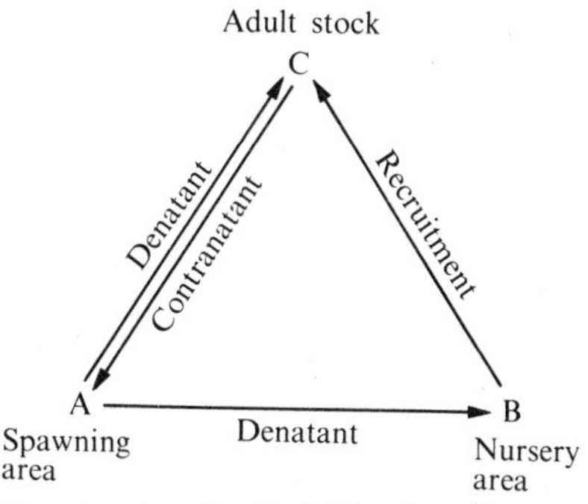

Fig. 2. The migration circuit of fishes in a diagrammatic form.

to feeding ground and back to spawning ground is not only a geographical description but also represents the transfer from generation to generation. It will be suggested below that the larval drift from spawning ground to nursery ground occurs in the same geographical position from year to year and also is the period when the major control of numbers takes place.

In temperate waters, the spawning ground tends to be fixed in time and space. Plaice in the southern North Sea spawn between the Thames and the Rhine and the peak date of spawning is 19 January, with a standard deviation of less than a week. Most of the arctonorwegian stock of cod spawns on the northern edge of the Vestfjord inside the Lofoten Islands in northern Norway. The peak date of catches has shifted slightly later during the last eighty years, but by less than a week. The Pacific salmon returns to its parent stream to spawn, and in nearly 50 streams or lakes for about 17 years the peak date of spawning had a standard deviation of less than a week. On much less evidence, much the same conclusion can be drawn from grab samples of herring eggs taken on the sea bed on the spawning grounds of the Norwegian herring. The Downs herring in the southern North Sea spawned on the same half a dozen very restricted places (300 m × 2 km) for more than a decade. Thus the positions and dates of peak spawning tend to be more or less fixed.

The spawning is, however, extensive in time: two to three months for the plaice and three to four months for the cod. The stocklets (or spawning groups) of salmon and herring do not spawn for very long, perhaps 10 days or three weeks, but, because each spawns at a slightly different time, the total spawning period for the stock is not very different from those of the cod and plaice. Within stocklets, the herring and salmon are quite densely concentrated, whereas the cod and plaice appear to be rather dispersed; for example there may be one plaice per 80 m on the spawning ground but one salmon per 5 m. The patch of plaice eggs and larvae is extensive, perhaps 30–50 km across, and it retains its spatial identity against the diffusive processes in the sea; if it were much smaller, 10 km across, it would disperse during the period of the larval drift. The general point is that despite the fixed date of peak spawning, eggs are laid for a considerable period over an extensive area; the long spawning season preempts the nursery ground from any competitor of the same species and the spread in space preserves the identity of the patch in the larval drift and so determines the position of the nursery ground.

In lower latitudes, perhaps those of less than 40°, there is no evidence of a fixed spawning season and not much of any fixed position. Egg distributions of the Californian sardine and anchovy are available for nine years and the standard deviation of the peak date of spawning is as much as three or four weeks. The positions of spawning are also variable as they probably shift with the points of upwelling up and down the coast. As upwelling depends on the wind directions and the angle it makes to the coast, the spawning grounds might be expected to shift with small changes in the wind direction. In the North Pacific subtropical anticyclone, the eggs and larvae of the species of tuna are distributed across the whole gyral, or oceanic current swirl, at nearly all seasons. There is a tendency to concentrate at the western end of the North Equatorial Current towards the Philippines, but it is less a spawning ground than an area of rather higher density of eggs and larvae of enormous extent–in millions of square miles.

There is a contrast between the fixity of spawning grounds in position and in season in temperate and high latitudes with the lack of any such regularity in low ones. In latitudes greater than 40° the production cycle stops in winter and rises again in spring or early summer to a peak of high amplitude in the production of carbon. In low latitudes the cycle is of low amplitude and it continues throughout all seasons; in the upwelling areas of subtropical and tropical seas there are variable seasons of high production but there are none with no production at all. In low latitudes food is available all the year round but poleward of 40° there is none until the production cycle has got under way.

The fixity of spawning ground and season in higher latitudes may be an adaptation to the variability of the temperate production cycle. The onset of production in temperate waters depends on the rising angle of the sun and the slackening stress of wind and the peak date can shift back or forth under different conditions. Then the success or failure of a cohort, or yearclass, may well depend upon the match or mismatch of the

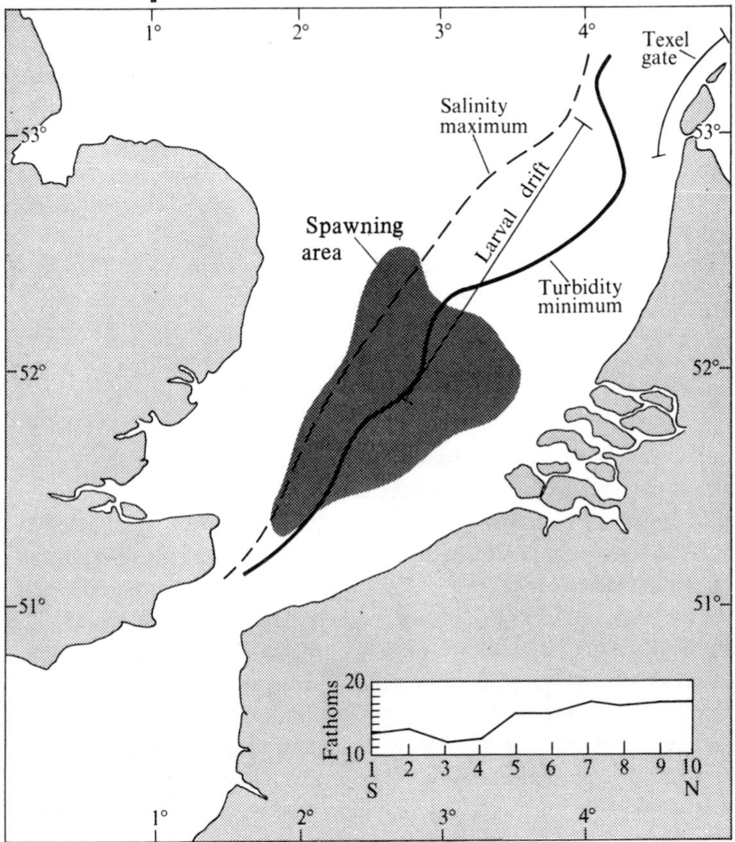

Fig. 3. The larval drift of the plaice from its spawning ground between the Thames and the Rhine to its nursery ground on the Waddensee inside Texel Island in northern Holland; the little figure below shows the depths from south to north along the larval drift.

timing of larval production to that of their food. Because the production cycle in low latitudes is continuous, food is available all the time and there perhaps is no need for tha adaptive mechanism found in high latitudes.

From their spawning ground between the Thames and the Rhine the larval plaice drift north to a point off Texel Island in northern Holland where the water flows inshore at certain states of the tide (Fig. 3). The nursery ground lies inside the western islands of the Waddensee. During their second, third, and fourth years the immature fish spread away towards deeper water and the migration can well be described as a diffusion outward from the Dutch coast (Beverton and Holt 1957). In the deeper water near the Dogger Bank, the maturing fish join the adult stock, and in the late autumn they migrate southwards to their spawning grounds. In each migration described the larval drift forms the geographical base to the circuit of migration in that both the spawning ground and nursery ground are more or less fixed. It is during this period that numbers are regulated in the main and the magnitude of the yearclass is determined. Presumably the position of the larval drift has been selected on an evolutionary scale to obtain the best environment for larval food with adult food accessible at the same time.

Each circuit depends on a current or current system. When the arctonorwegian cod are drifted north in the West Spitzbergen current they are carried along the western edge of the Svalbard Shelf. They then reach the top of the shelf, possibly by their daily vertical migration or possibly by local twists of the current, for example into the North West Gully, north of Bear Island. They then disperse across the shelf, having left the west Spitzbergen current. However, they do not penetrate very far into the cold arctic water masses to the north and east; very roughly the boundary of their feeding grounds in the Barents Sea is indicated by the 2°C isotherm on the bottom. Later in the year they move into deep water and migrate south to the Vestfjord either against the current or with the counter-current and leave it when they round the Lofoten Islands into the Vestfjord. Not only does the stock make use of the current system, so to speak, to maintain its migration circuit, but the stock may be said to be contained within a current structure. For example, the Norwegian herring are contained within the main gyral in the Norwegian Sea and the North Pacific albacore are contained within the North Pacific subtropical anticyclone, that is, between America and Japan and between the equator and 40°N.

Most of the abundant commercial species of fish appear to be migratory. As fish grow up through the marine ecosystem so they exploit it in different areas during their lives. Larval plaice feed on *Oikopleura* and copepod nauplii (young stages of small crustacea) in the Southern Bight. Immature plaice feed on sand crustacea and little worms in the Waddensee and just off the coast. Adult plaice feed on bigger worms in the deeper water around the Dogger Bank. Adult growth may not be food-limited to any great degree so it does not matter where they feed. The larval growth may be sharply food-limited in that they have to grow to avoid death. They grow at about 6 per cent per day and die at 5 per cent per day. To maintain the stock, the balance between growth and death during the larval drift generates the later year classes and the numbers are regulated in the same processes. (Year class, or cohort, means a brood spawned in a given year). The position of the larval drift from spawning ground to nursery ground is probably the best for the stock to maintain itself in the face of environmental change in an evolutionary sense. Then if the adults can return to the spawning ground by using the current system, the abundance of the stock depends on the productivity of the water along the course of the larval drift and upon the competition with other carnivores.

3. The unity of the stocks

A stock is a large population of fishes distinct from its neighbours and differences between stocks should be detectable genetically. Within stocks mating should be randomly distributed, but if a stock is subdivided into groups, or stocklets, between which material is exchanged from generation to generation, the variation in the stock is increased. There are three or four spawning grounds of plaice in the southern North Sea; of fish tagged on them, nearly all return to the same place in the following year and do not stray to another. Between the fish found on two grounds there are differences in otolith structure that suggest an origin very early in the life cycle, perhaps on the nursery ground. If there is exchange between the plaice stocklets it may occur by displacements in the larval drift. Of the Pacific salmon that return to their native grounds, a proportion, albeit very small indeed (3 per cent), strays to other streams. The British Columbian herring spawns on the shores of Vancouver Island and the shoreline has been divided into statistical areas; a large-scale tagging experiment showed that the stray from one statistical area to another decreased with distance apart, as might be expected, but also with age (Harden Jones 1968). The discrete and restricted spawning grounds of the herring, both in the southern North Sea and off the coast of Norway, suggest a similar structure of stocklets within a stock, but there is as yet no evidence of stray or structural difference between them.

In the past, differences between stocks have often been established by counting the numbers of vertebrae, fin rays, and so on. Such differences are environmentally determined at a very early age and so the method was supposed to indicate differences between spawning groups. However, the temperature differences, for example, between adjacent spawning grounds could well be small and the differences in the measured characters would be small. Inevitably there must appear stocks or stocklets that were distinguished where environmental differences were reduced to very low levels. Today the blood proteins are used to establish differences between stocks because they are genetically distinguishable and the samples are easy to take. The common proteins are the haemoglobins and transferrins and they are found in cattle as well as fish. A large amount of observational and experimental work has been carried out on cattle, particularly in the development of family characteristics; so the genetic argument based on cattle experience can be extended to fish.

The study of blood proteins in fish started in the United States where some herring stocks were distinguished on immunogenetic grounds. Three subpopulations were discovered in the California sardine stock and a number of differences were established between groups of tuna in the eastern tropical Pacific. In the Atlantic the most successful application of the method has been the establishment of genetic differences between the different cod stocks by the European biologists (Fig. 4). Such differences are estimated in a (2×2) table. In the cod stocks there are three haemoglobin and seven transferrin alleles. Differences are estimated in a $(2 \times n)$ table; between one stock and another there may be only one chance in 10^4 of mixture. The study of the populations themselves becomes very difficult if there is significant mixture between the groups examined. Particularly in the study of the dependence of recruitment on parent stock is it necessary to establish the unity of the stock, because the recruitment is variable and any components of admixture should be excluded.

The genetic studies yield other forms of information. For example, in the North Sea, tagged cod appear to segregate in different groups, but no genetic differences has yet been established between them. There are distinct spawning grounds for example in the Southern Bight and on the Flamborough Off Ground. Perhaps there is a North Sea stock that comprises a number of stocklets. Another point is that there is a trend in the blood proteins across the North Atlantic which indicates that the stock

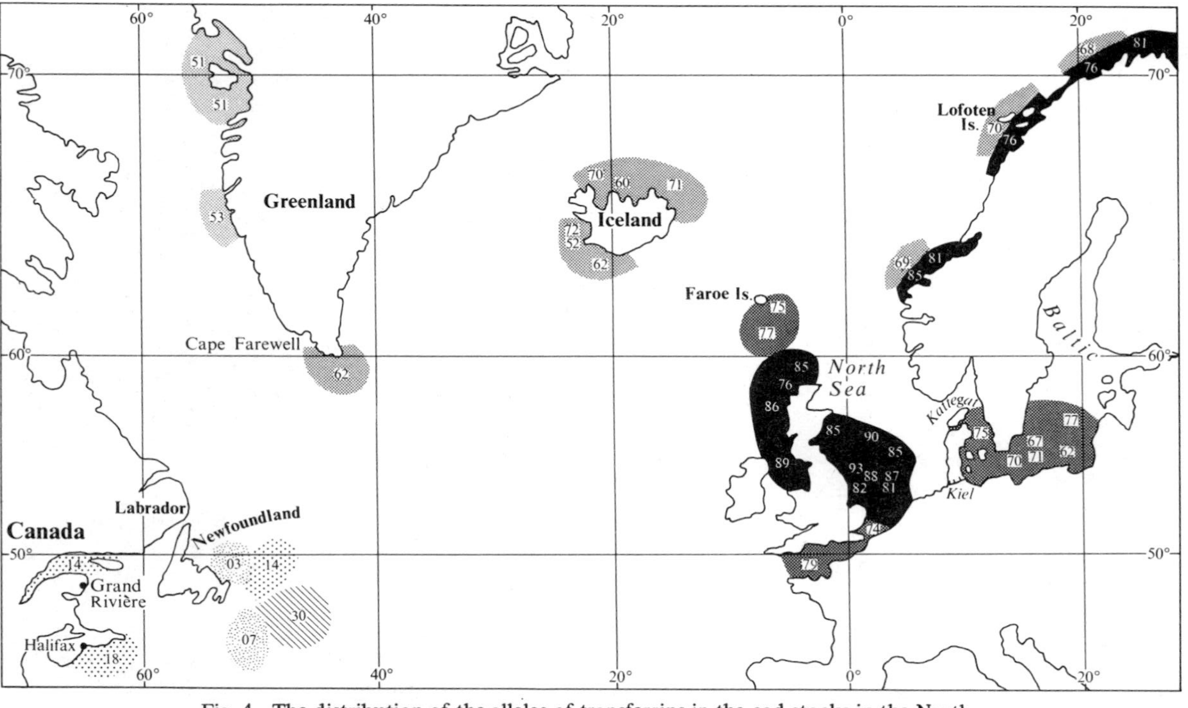

Fig. 4. The distribution of the alleles of transferrins in the cod stocks in the North Atlantic.

on the Grand Banks and those in the eastern ocean have been separated for a very long time. Similarly the Baltic cod stock is very markedly different from that in the North Sea; yet when there is a year of high salinity inflow, which occurs from time to time, North Sea cod penetrate the Baltic Sea. In the Vestfjord differences can be established between the arctonorwegian cod and the fjord cod that spawn in the same area. The question is then raised how the populations segregate from each other and it resembles the question of how some fish find the same restricted spawning grounds year after year.

Stock unity is maintained by reproductive isolation from generation to generation. The stock is contained within a particular current structure which the fish can board or disembark from at particular points. Within each stock there are smaller groups or stocklets that appear to be maintained by the capacity of the fish to home on a particular and restricted ground. There is exchange between stocklets, presumably caused by stray from homing. There is no such exchange between stocks because the stray is very small and reproductive isolation is secure. Thus the study of migration is also that of stock unity, not only in broad terms, but also in detail.

4. Conclusion

This brief account of the natural history of some fishes of commercial importance is one restricted to the need to understand the stock dynamics described in the next chapter and the nature of the fisheries that exploit those stocks. The migration circuit provides a basis of stock unity which can now be tested by genetic methods. It also provides a *raison d'être* for the regularity of fisheries that is observed in time and space. For example, the arctonorwegian cod have been caught on the northern side of the Vestfjord in February and March for centuries; reliable records go back to the twelfth century. Recovering spent fish are caught in June and July on the Svalbard Shelf in the Barents Sea and on Skolpen Bank (east of North Cape) regularly each year. In the autumn and winter large cod are caught on the banks north of Vestfjord as they migrate south to the spawning ground. Such regular fisheries can be described and explained in terms of the migration circuit. Sampling systems in the ports can be established to record information regularly from such fisheries as part of the exploitation of one stock.

One of the most important questions at the present time is that of the dependence of recruitment on parent stock. The nature of the dependence is only dimly understood, but it would not be elucidated unless the unity of the parent stock were well established. More information is needed on the growth rates and death rates of larval fish and of the processes by which the increase in cohort biomass established. Fisheries biologists have tended to examine the adult age groups most thoroughly and to pay little attention to life during the larval drift or on the nursery ground. Spawning grounds and nursery grounds need often to be described in some detail and the nature of the larval drift between them needs to be established. The larval and juvenile natural history of many species of fish remains to be well described.

3
The dynamics of fish populations

1. Introduction

Any population of animals is maintained by the balance between its birth rate and death rate. The birth rate is probably determined by natural selection (Lack 1954), but the death rate expresses the capacity of the environment to carry the population. Animals die through predation, old age, parasitism, or disease, although the proximate cause of death may always be predation, particularly in the sea. Deevey has published the life tables of a number of animals in widely different groups that showed the trends of mortality with age. Common to all was the decline in mortality with age during juvenile and immature life with a least death rate in the early and middle mature age groups and an increase in late middle age and old age. Such a trend occurs in fishes, but the early mortalities during the larval drift are very high, 5–10 per cent per day, whereas in early and middle maturity they are reduced to 5–20 per cent per year. As indicated in the last chapter the decline of mortality with age describes the progress of the cohort through the ecosystem and the magnitude of mortality at a given age depends on the degree of predation. If a fish population is transferred to a new ecosystem, the birth rate and death rate may not be matched. For example, striped bass were released in San Francisco Bay in 1880 in small numbers and twenty years later a catch of 500 tons per year was taken. Because the birth rate exceeded the death rate, the ecosystem could carry the extra stock and the striped bass were able to colonize San Francisco Bay. The difference between the rates is exponential and the population increases geometrically with time until the rates balance and the numbers in the population stabilize. Most fish populations colonized their migration circuits long ago and the system remains in balance until an additional death rate is imposed by fishing.

There are three characteristics of fish populations. First, they live a long time—for decades in some cases; secondly, they can be readily aged, at least in temperate waters; thirdly, a fair proportion of the stock is counted and weighed on the quays and markets. In contrast to insect populations, they are not transient nor are they quite so vulnerable to climatic variability. In contrast to mammal populations, they are easily aged and in some places they are readily counted. When age determination is well established, growth rates and death rates (due to fishing and to natural causes) can be readily estimated. The death rate by fishing can be estimated by tagging a number of fish, as indicated above, and the proportion recaptured indicates a rough measure of fishing rate, provided that fishing vessels are distributed over the expected area of recapture. By far the most important point is that in a well-exploited stock up to one-third or one-half of it is recorded as catch on the fish market and much of this fraction can be measured in length. Such records extend back for up to fifty years (and sometimes more) for a few fish stocks. No other wild and large populations of animals are so well documented as some stocks of fish.

2. Market sampling and stock density

Catches are recorded by position of capture and often by the time spent fishing, which is also called the fishing effort. Then the catch per unit of fishing effort f, for example catch per day's absence, or catch per 100 hours trawling, or catch per purse seine shot is an index of stock.

In a tagging experiment, tagged fish are released (the stock of tags) into the sea and are recaptured by fishing vessels (the catch of tags). The ratio of catch of tags to stock of tags is the instantaneous fishing mortality, F, i.e. catch/stock $= F$; so stock $=$ catch/F. $F = qf$, where q is the catchability coefficient, and stock $=$ catch/qf.

Let N_0 be the stock in numbers at the beginning of the year and N_1 that at the end. Then $(N_0 - N_1)$ is the number of deaths during the year.

Z is the total instantaneous mortality, which is the sum of fishing mortality and natural mortality.

Then, the catch equation can be written:

$$\text{Catch} = (F/Z)(N_0 - N_1)$$

$$\text{Then Catch}/F = (N_0 - N_1)/Z$$

The left-hand term is the stock, as shown above. The relation between stock and stock density is given by:

$$\text{Catch}/F = (1/q)(\text{Catch}/f).$$

The left-hand term is the stock and the right-hand one stock density (because fishing mortality is proportional to the time spent fishing). To the fisherman it is the average catch and therefore an index of profit. Hence, when his catches cease to be profitable the stock density has probably decreased.

The catch of a single vessel or a group of vessels may be expressed in stock density. On English markets fish are measured in length and some earstones or otoliths are collected; from the annual rings on the earstones (or scales in herring and salmon) the age can be determined. An age–length key is constructed and a large number of length measurements from a group of vessels can be converted into an age distribution in stock density. Such individual distributions are grouped by ports, areas, and months into annual distributions of age in stock density, which are in the same units from year to year.

The importance of this technique is that growth rates and mortality rates can be estimated directly from the age distributions. The mortality rate is the ratio of numbers in a cohort or year class in successive ages and the growth rate is the ratio of weights (or lengths) in successive ages. But to solve the catch equation we need to know the fishing mortality. It can be obtained in a tagging experiment as noticed above or by relating total mortality to fishing effort in a period of changing effort. However neither method is quite straightforward. The first relies on the assumption that the tagged population is thoroughly mixed with the untagged one, which might take some time in a large stock. The second relies on the assumption that natural mortality is constant with age and that effort is reliably measured during the period. However, with care, reasonable estimates can be obtained.

Although stock density is a true index of stock it can become biased if catches vary because the stock is only partly sampled by the fleets. The bias might be detected directly by sampling with research vessels outside the normal range of the commercial fleet. It can also be evaded to some extent by avoiding the dependence of total mortality on effort and by estimating fishing mortality in the catch equations as the ratio of catch to stock. Recent developments with this form of technique have led to quite accurate simulations of the trends in catches with time.

A good system of market sampling requires that the quantity of each haul be recorded by its position of capture. An ideal system would demand that the length measurements on the quay should be identifiable to the same positions of capture, but this can only rarely be obtained. Many years ago, a fishing vessel would go to a single ground and work for a period; the total catch, with the number of hauls made on that ground, would indicate the average catch by position of capture. Today, the vessels shift grounds in the North Sea within trips as changes in prices are reported on the radio; and on a broader scale I have seen trawlers off South Africa haul their gear to steam to the grounds off the River Plate. Under such conditions, where the fleets are mobile, each haul should be recorded by its position of capture; nor would it be difficult for the fishermen to tell the fish measurers which fish came from each ground.

The good system of sampling and fish stock on the quay or on the market provides the fisheries biologist with the essentials of his science. The quality of the science depends on the quality of the sampling. Long ago, proud skippers were unwilling to yield what they considered were trade secrets to anybody. As the fleets become more international and more mobile many skippers were persuaded that conservation is badly needed and that the information should be given freely and in detail to the fisheries biologists.

3. The description of fish populations

The first fisheries biologists at the turn of the century expressed two principles. The first was that the catch had to be balanced by the reproductive capacity of the stock; not only are the birth rate and death rates of the stock approximately equal in the unexploited state but the birth rate should increase relatively to compensate the loss of stock in catches. The second principle was that because most bottom-living fish grow by an order of magnitude during their adult lives, fishermen would lose money if they caught them when they were too small. Both Garstang and Petersen separately enunciated the principles that we now describe as the conservation of recruitment and the conservation of growth. If, through catching fish before they grow much growth is not conserved, a stock might suffer from growth overfishing; and if recruitment is not conserved it might suffer from recruitment overfishing because stock is reduced so far that recruitment declines absolutely. However, these principles were not put into practice because the early fisheries biologists found the variability from haul to haul of a North Sea trawler so high as to be unmanageable. Their solution to the problem of growth conservation was to consider the transplantation of plaice from the continental coasts of Europe, where they were crowded and small in size, to richer feeding grounds on the Dogger Bank. This attractive idea was to remain a valid aim of fisheries biologists until the 1950s, when it was shown that the added increment of growth of plaice transplanted from the Baltic to the

Limfjord (in western Denmark) did not meet the cost of transplantation. Again, plaice transplanted to the Dogger move off it into deeper waters, which is perhaps why the feeding grounds appear to be so rich.

The next stage in the description of fish populations was the discovery of the inverse relationship between stock density and fishing effort. It was successfully applied by W. F. Thompson to the Pacific halibut fishery. With the use of an early form of the catch equation he was able to show how a decrease in fishing effort would increase stock density and hence stock, and then catch. An important part of his method was to proceed slowly year by year as annual observations of stock density and of fishing effort became available. When the winter halibut grounds were closed in 1930 he was able to proceed in subsequent years in such a self-adjusting manner.

More formally, animal populations has been described by the logistic curve which stated that the net rate of increase (from generation to generation) was balanced by a constant assumed to be the carrying capacity of the environment. Estimates of the constants were obtained from the differences in the population from year to year. The colonization of San Francisco Bay by the striped bass could be well described by such a curve. During the twenty-year build-up of the stock the net rate of increase was always greater than the modifying constant so the population continued to rise. When the carrying capacity of the environment was saturated, the two constants would be equal and the population would then remain in a steady state as far as possible. Hjort *et al.* introduced the logistic curve to the analysis of whale populations in 1933 and Graham (1935) applied it to the southern North Sea stocks of plaice. It follows from the logistic curve that if the stock is reduced by fishing the carrying capacity of the environment is increased and so the net rate of increase rises. Thus, as fishing mortality increases, the recruitment to the stock should either rise or remain constant, but in any case it would rise relatively to stock. Both Graham and W. F. Thompson (who also discussed the point) recognized that at much reduced stock the recruitment must decline and that at zero stock there must be zero recruitment. However both thought that such a reduction in stock would not be obtained by ordinary levels of fishing. Graham was able to show that there was an optimum catch, for with no reduction in recruitment with fishing, the problem is to obtain the maximum increment of growth in face of the loss of numbers. In other words he faced the same problem as Petersen: to conserve growth.

The method was extended by Schaefer. With an estimate of fishing mortality the dependence of stock density on fishing effort was converted into a parabola relating catch to stock and the maximum sustained yield was obtained at half the virgin stock. Catch can similarly be related to fishing effort and it is a very simple convincing relationship, provided there are enough data. Confidence limits can be put to the curves and after a suitable number of years the maximum sustainable yield can be reasonably well established, provided that the fishery develops slowly enough (for example, in Fig. 7 p. 46). The disadvantage of the method is that the observations are acquired annually and it make take a decade or so to establish the position of the maximum.

Another disadvantage of the method is the fact that the vital parameters of recruitment, growth, natural death, and death by fishing are confounded. In a stock in which the recruitment is greatest at or near the point of maximum sustainable yield, little harm is done by confounding the parameters. If the maximum shifted in magnitude, it would be desirable to know if the change were due to recruitment, to natural death, or to growth. However, as knowledge advances it seems unlikely that great stock changes occur due to density-dependent effects in growth or mortality among the adults. The most probable source of change is then that due to recruitment. If the stock can be aged, then the recruitment changes can be shown. The present application of the Schaefer model is to stocks like the yellowfin tuna in the eastern tropical Pacific that cannot be readily aged.

Summarizing, the descriptive model or the Graham-Schaefer model can be used to define the maximum sustainable yield if time permits. But if decisions in management are needed quickly, as in these days of mobile fleets, the descriptive model cannot give the answer in time. The maximum yield would be well estimated by this model and the stock should not suffer from recruitment overfishing, that is, the stock should not be reduced so low that the recruitment itself is impaired. Perhaps the most valuable use of the descriptive model was to establish the maximum sustainable yield for the antarctic blue whale (Anon 1964); the apparently simple presentation of the facts eventually convinced the International Whaling Commission that conservation was needed and needed quickly.

4. The analytic model

The analytic model was first formulated by Russell, who said that any change of stock from year to year comprised increments of growth and recruitment and decrements of mortality (natural deaths and deaths by fishing). The population is considered as the result of changes in the vital parameters rather than as merely a series of increments or decrements in weight or numbers. The parameters are evaluated independently of the population and a model can be synthesized from them that simulates the behaviour of the stock under different conditions.

The analytic model was first used by Beverton and Holt (1957), but they also made other innovations. They considered stock as the integral of numbers and of weight from the age of recruitment to the age of extinction in a cohort or year class. Hence the age structure of the population was all-important, but it was made available in the market sampling system in use in England and Wales. Growth and total mortality were estimated directly from the sampling system and additional information was obtained on fishing mortality from tagging experiments.

In order to integrate numbers in age, the rate of change of numbers was attributed to mortality, fishing and natural. This is straightforward but the growth of fishes had to be formulated in a manageable way. The von Bertalanffy growth equation was used, which states that growth in length or weight tends towards a maximum at infinite age. By fitting the curve to data, two constants are extracted, the maximum length or weight and the rate at which the maximum is attained in age. The integration of stock in numbers and in weight in age incorporates the two mortality coefficients and the two growth constants. Catch is the product of stock and fishing mortality and so it is quite easy to express catch as a function of fishing mortality. By far the most important point is the capacity to do it quickly once the constants of growth and mortality have been established.

The growth of fishes has been described in a number of equations. Choice depends as much upon the purpose to which the results are to be put as upon the theoretical bases or the statistical precision obtained from the data. On theoretical grounds there is little to choose between them and most fit the data as well or as badly as the others. The von Bertalanffy equation does describe the continued growth of fish throughout adult life (which may amount to an order of magnitude or more), the characteristic that led to the belief that they were potentially immortal. Further the observations of length or weight can be graduated quite accurately so long as growth can be considered to be isometric; it is very hard to describe the transition from a larval fish to a metamorphosed one when the relative dimensions can change radically. The most important point about the von Bertalannfy equation is the facility with which it can be integrated in age in conjunction with the mortality rates.

The yield curve of catch on fishing mortality for the southern North Sea plaice showed that the maximum was found at about one-third of the fishing mortality observed during the 1930s. A gain of about 20 per cent in total catch and of about a factor of three in catch per unit of effort could have been obtained by putting two-thirds of the fishermen out of employment (Fig. 5).

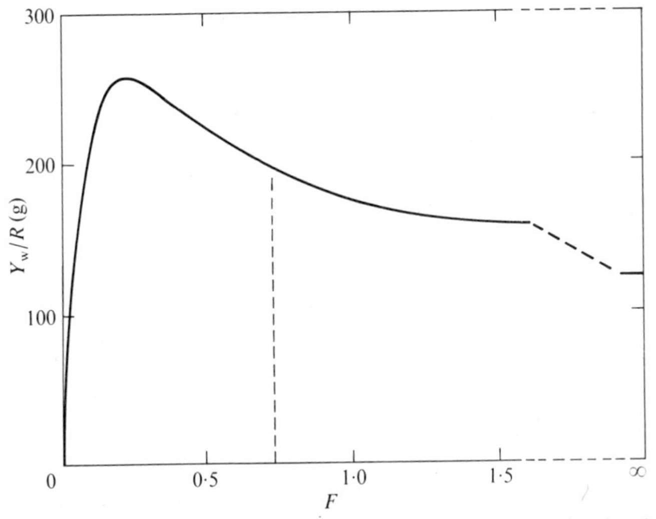

Fig. 5. The dependence of yield per recruit on fishing mortality for the plaice of the southern North Sea; the ordinate at $F = 0{\cdot}73$ represents the state during the 1930s (Beverton and Holt 1957).

However in Britain during the 1930s a form of mesh regulation was devized by inserting larger meshes in the cod end of the trawls in order to allow the little fish to escape. By increasing the mesh size, the effective age of entry to the fishery could be raised. Then the maximum yield would be obtained at a higher rate of fishing mortality. With the use of such models it was shown that considerable gains could be made by increasing the mesh size, or age at first capture, without having to reduce the numbers of fishermen.

The models were extended by Beverton and Holt in two important ways. First they considered the possibilities that growth and natural mortality were density-dependent during adult life. There was little evidence for this but the theory of balance required it; that is, if recruitment were to rise under the pressure of fishing, the cause must be a density-dependent one; and if recruitment were density-dependent, then growth and natural mortality were probably as a consequence also density-dependent. A number of ancillary models were constructed which showed how far such density-dependent effects, if they existed, might alter decisions in management based on the simple curve of yield on fishing mortality. However, in recent years, people have tended to believe that density-dependent processes become less important in the adult age groups.

Strictly the yield was not expressed as catch but as yield per recruit. If recruitment were considered to be constant in stock then a yield-per-recruit model expresses what happens in the real world; in the next section the relation of recruitment to parent stock will be considered in more detail. The second extension of model techniques used by Beverton and Holt was to combine a stock–recruitment curve with their earlier model to make a self-regenerating yield curve. With such a model one might simulate the whole behaviour of the stock from the independently determined parameters. This particular model was not pursued very far because the relationship between recruitment and parent stock was considered to be too variable.

Indeed the relationships available in the early 1950s were so scattered that Beverton and Holt reached a conclusion that was slightly different from Graham's. Graham believed that as stock reduced under the pressure of fishing, recruitment increased relative to stock (or even absolutely), and he concluded that fishing would never reduce the stock far enough to reduce recruitment either relatively or absolutely. Beverton and Holt thought that the variability of recruitment was so high that the yield-per-recruit models could be applied exclusively. However, the same variability masked any true decline of recruitment with increased fishing until it was too late, particularly amongst clupeid stocks.

Thus there is a sense in which the yield per recruit model solved the problem of growth overfishing and at the same time inadvertently generated that of recruitment overfishing. As will be shown below, it happened because the nature of the stock–recruitment relation in herring-like fishes or clupeids (and to some extent in cod-like fishes or gadoids) was not understood. Because of the recent incidence of recruitment overfishing, there might be a tendency to revert to the descriptive models. The failure of the yield-per-recruit model in some instances is really a historical accident because information was lacking. The scientific advance made by Beverton and Holt was considerable in that they exploited the age structure of the population as shown in the market sampling system; and they devized models synthesized from the vital parameters of the population on the basis of which management could take decisions quickly rather than after a decade or so of waiting. The real point is that fisheries biologists should not allow themselves to use other than analytic models in the future, unless they are forced to do so. Management needs clear information and the form in which Beverton and Holt put it is that needed.

5. The dependence of recruitment on parent stock

As noted in an earlier section, the use of the logistic curve to describe the dynamics of a fish stock implies an intuitive understanding of the dependence of recruitment on parent stock. At a middle range of stock, recruitment may often be described as being independent of stock. Therefore in the mortality between egg and recruitment there must be a density-dependent component. Such a phrase almost presupposes that another part of the mortality between egg and recruitment is independent of density. Most fisheries biologists would consider that a large proportion of this mortality must be independent of density because much of the variability of recruitment is determined environmentally.

The problem is how the density dependent mortality is generated. There are at present two solutions. The first, due to Ricker and Foerster suggest that there was a critical period in larval life during which predatory mortality depend upon the time to grow through it; if growth was slow the mortality lasted for longer than if the growth was quick. Mortality then depends on the food available, and because food is shared among increasing numbers, the death rate is also density-dependent. Later, Ricker (1958) developed another solution by which mortality was generated by the aggregation of predators (or even cannibals of older year classes) on the stock of eggs or larvae. The density-dependent mortality was generated by differences in stock from generation to generation. Beverton and Holt (1957) subsequently pointed out that the same effect could be achieved by a mortality dependent on available food, so long as the critical period occurred early enough in the life cycle to accommodate the differences in stock from generation to generation.

Ricker's curve describes the dependence of recruitment on parent stock as being governed by the aggregation. That of Beverton and Holt describes it as being governed by the availability of food as well as by the density-independent mortality. In one form the two equations are identical, although derived in different ways. As they are used in a descriptive manner with averaged constants, the difference is of little practical importance, although of considerable theoretical interest. When fitted to observations such a curve may be nearly linear, flat-topped, or markedly domed in stock. in the first case, recruitment increases with stock to a maximum; in the second it is independent of stock over an extensive range of stock; and in the third it decreases at high values of stock. Beverton and Holt developed another stock and recruitment relationship which sets density-dependent mortality as a function of numbers and the curve of recruitment is asymptotic to stock, which may not fit some sets of observations, particularly those of cod-like fishes. This suggests that density-dependent mortality in itself is not enough to explain the observations; the system of Ricker and Foerster implies that the growth rate also plays a part, particularly if it is also density-dependent.

The question arises whether the nature of the stock–recruitment curve differs in the various groups of fishes. Cushing and Harris have suggested that density dependence can be related to fecundity, in fact to the cube root of fecundity or the distance apart of the larvae in the sea. As density dependence increases one might imagine the curve of recruitment on parent stock bending from a near linear form to a flat-topped one and finally to a dome. This progression is one from herring to plaice and then to cod, which is an increase in size and hence fecundity (because fecundity is a function of weight).

Any stock and recruitment curve cuts the bisector at the point of stabilization, about which the virgin or unexploited stock varies. A low recruitment reduces stock below the point of stabilization, which generates a higher recruitment, and so stock is brought back to the stabilization point; above that point in stock, the same process works in the opposite direction (Fig. 6). Hence the stock stabilizes itself about this point. If the stock and recruitment curve for the herring is nearly linear, or perhaps asymptotic, there is little capacity for stabilization. That for the cod may be markedly dome-shaped indicating great stabilizing capacity. The cod would then be expected to withstand a greater degree of environmental change than the herring. Since the twelfth century the arctonorwegian cod stock has been fished in the Vestfjord continuously, albeit with some secular variations in the catches. During the same period, the Norwegian herring fishery was extinguished on a number of occasions as the climate changed in a periodic manner.

Just as the herring-like fishes may be vulnerable to secular climatic changes, so are they sensitive to excessive fishing. Herring do not grow much during their adult lives and there is no point in conserving their growth. But as fishing increases on a herring stock, recruitment may tend to decline and, in the extreme, recruitment overfishing may supervene. A number of large stocks of herring-like fishes have declined during periods when the fishing effort has remained high; indeed many of the

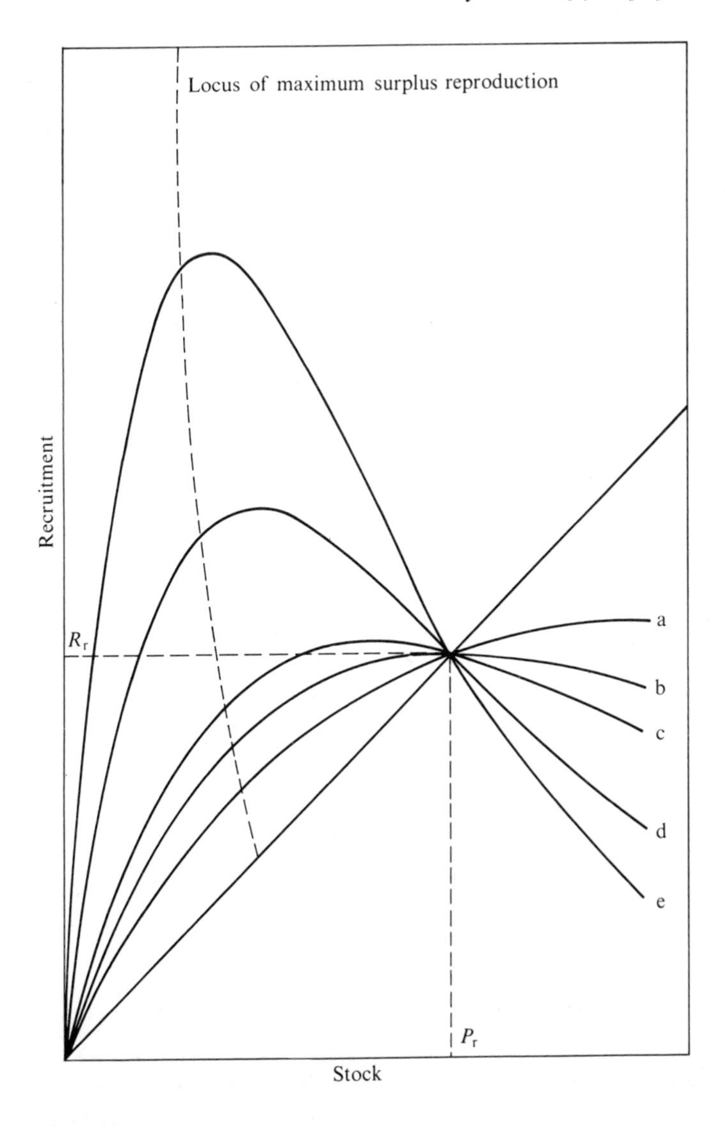

Fig. 6. The dependence of recruitment upon parent stock (Ricker 1958). The family of curves represent different degrees of density dependence; the sharper the dome the greater the density dependence. The pronounced dome tends to be characteristic of the cod-like fishes and the less pronounced one of the herring-like fishes.

herring fisheries in the north-east Atlantic have now vanished. If cod stocks resist environmental change more effectively than do herring stocks, it is likely that they also resist the pressures of fishing to a greater degree. Indeed, few if any demersal (sea-bottom) fisheries have been extinguished when the fishing effort was intense, which is in itself a contrast with the herring fisheries.

The most important part of the stock and recruitment problem is to devize a yield curve that incorporates the average changes of recruitment with changes in stock. Beverton and Holt did this in their self-regenerating yield curve, but they used the asymptotic form of the stock–recruitment relationship that may no longer be valid. Further the very high variability of the recruitment cannot be ignored, even if some of the curves can now be fairly well described in statistical terms. The great virture of the yield-per-recruit formulation is that models can be built from parameters about which the variability is not very great. Any stock–recruitment model at the present time would incorporate one or two highly variable components. Another great disadvantage is that the stock–recruitment curve is slowly established by decades of annual observations and thus shares all the drawbacks of the old descriptive models. To summarize, an analytical form of yield curve is needed which accounts for the variability of recruitment, yet evades it in presentation.

6. The present state

The application of population dynamics to fisheries has a long history. In recent years, however, with the proliferation of international commissions it is used in practically every major fishery in the world. All the methods described in this chapter continue to be used in one form or another and it is in such a combination of methods that the science reaches maturity.

One valuable recent development arises from the catch equation. Usually fishing mortality is assumed to be related to the time spent fishing or the fishing effort by the coefficient of catchability. However, fishing mortality can also be estimated from the ratio of catch to stock, and this emerges directly from the catch equation. The sum of catches throughout the year class is called the *virtual population* and it has some value because it is the least estimate of stock. In a new modification to the method with estimates of natural mortality and of total mortality (Gulland in Garrod 1967), the stock of the oldest age group is estimated from the ratio of catch to total mortality. Then in the penultimate age group, stock is estimated as an increment of the last group by the total mortality expressed positively. From the ratio of catch to stock in the penultimate age group a new estimate of fishing mortality emerges. In this way stock is elaborated back through the year class and yearly estimates of fishing mortality emerge. The method requires that an estimate of natural mortality is obtained independently. There are two considerable advantages to this method; first that much of the variability in catch is reduced and secondly that the stock can be expressed in absolute numbers. Since about 1971, catch quotes have been introduced in the North Atlantic as a mode of çonservation and there are obvious advantages in putting catch in absolute numbers rather than in stock density. This method of cohort analysis has considerable advantages in long-lived stocks, for in the abundant age groups that contribute to the catches the estimates of fishing mortality are good.

The use of models has developed in a particular way. People have tinkered with the descriptive Schaefer model and with the yield per recruit model. The new development, however, is in use of simulation models. For example Rothschild and Balsiger have created a simulation of the Bristol Bay salmon fishery in Alaska, in which an attempt is made to forecast the timing and quantity of the run as it approaches the rivers. Such information is useful to the economic management of the fishery. Clayden used the basic inverse relationship between stock density and fishing effort in the North Atlantic to model more complex relationships of the distribution of fishing according to the relative magnitudes of stock density and distances from port; in some regions trends of stock density and of catch were well simulated in hindsight and are now used for prediction. As fisheries biologists become more familiar with such methods, simulation models will proliferate.

The basic methods of fisheries research depend on the catches, and there is sometimes the need to estimate stock independently of the commercial fleets. The number of eggs in the sea divided by the fecundity estimates the stock of spawning females. This method can work only during the spawning season, and, from the nature of the migration circuit, fisheries occur in places other than on the spawning ground. Forms of acoustic survey are being developed by which fishes can be sized roughly and counted. Fish cannot yet be identified acoustically but the signal from a single fish is approximately related to its volume and an average size can be estimated. But samples of fish must be caught to identify them, and so any acoustic survey must be linked to a survey by capture. In a given area it might be possible to use fifty trawl hauls, but the numbers of samples of acoustic density could amount to millions. Thus the real purpose of an acoustic survey is to increase sampling power considerably. It is also possible that an acoustic density is more reliable than a trawl density because the quirks of behaviour in capture are evaded. The stock of hake off South West and South Africa was estimated acoustically, as was the stock of spawning cod in the Vestjord in northern Norway and the stock of sockeye salmon smolts in Lake Washington in Seattle; the latter survey was remarkable in that the error on the stock estimate was as low as 10 per cent. Of the two survey methods independent of catches, egg surveys are perhaps variable and are limited to spawning grounds but yield estimates of total adult stock, whereas acoustic methods are potentially more accurate and can be used anywhere, including spawning grounds.

The most pressing scientific problem facing fisheries biologists is that of the dependence of recruitment on parent stock. There is not enough information on the location of spawning grounds and nursery grounds, although in a few stocks they are very well known. Not enough is known of the growth rates and death rates of fish larvae, for only in a few instances have they been sampled right through the period of the larval drift. The same may be said of the little fish on the nursery grounds although a fair amount of knowledge is being accumulated on the flatfish and on the Pacific salmon. Nothing is known of the dependence of larval mortality on food and hence on density although the possible processes have been simulated. In temperate waters, it has been suggested that larval fish grow up with their private food supplies because their production is timed approximately to that of their food. Differences in food from year to year are perhaps related to differences in the magnitude of recruitment between years. There is a single process, the exploitation of the available food, in which numbers are regulated and the magnitude of recruitment mainly determined. The real problem is to disentangle the two facets of the same process at sea.

7. Conclusion

The science of fish population dynamics is a developing one. In the early stages, particular solutions were proposed as ultimate ones. Later, each of the early solutions became fallible in one way or another and today there are a number of solutions and methods that can be partially combined with each other. The object is always to assess how much can best be taken from a given stock with a view to international regulation. In the past, stocks have been controlled by minimum landed sizes, by mesh regulation, or by closed areas or seasons, all of which are rough but effective means of regulation. Recently catch quotas have been introduced. The problem used to be how to obtain a maximum or optimum yield on average, but with catch quotas the new problem is how to predict catches. The consequence of this change of outlook is to demand much more precision in assessment and to place greater strain on the methods of data collection and on the basic biological material.

In some senses the information collected on the markets is insufficient and not enough use is made of the material already collected. The statistics of the collected material have not yet been fully exploited nor are they processed rapidly enough. For example, the variability of growth and death within the system of data collection have been investigated only in the most summary manner.

4

The first steps in conservation

1. Introduction

Men have always known in principle that stocks of fish needed conservation but they have also hoped that their catches did not really harm the stock. Further, it has always been difficult to take the step in imagination from the relatively small catch by a single boat in the vast expanse the ocean to the immense catches made by the international fleets from a single stock, however large. If the fisherman sees the catch by his gear decline over the seasons, the cause to him is necessarily elsewhere and is often attributed to another gear that supposedly kills more fish than are needed. In the mid-nineteenth century pilchards were caught by very large shore seines on the coasts of Cornwall in south-west Britain. The seine net fishermen complained bitterly that their catches had fallen since the introduction of drift nets for pilchards. This complaint was almost certainly well based because the drift nets were shot in the path taken by the fish round the bays before they reached the seine nets and so deprived them of local shoals. However, it is very unlikely that the stock of pilchards was affected by the catches taken at the time. As an economic consequence the drift netters replaced the shore seiners and now they themselves are giving way to midwater trawlers.

When the herring stocks declined in the southern North Sea under the pressure of fishing, the driftermen that worked off the East Anglian coasts blamed the trawlers that exploited the spawning grounds further south. In the long term of history they were right in so far as the trawlers entered the fishery long after the drifters; in fact the trawlers started work in about 1950, whereas driftermen had shot their nets in the southern North Sea for centuries. However aggrieved were the drifter skippers, the trawlermen were free to enter the fishery and drift nets still killed herring, if perhaps less efficiently than trawls. In Cornwall the replacement of shore seiners by drift netters took place purely on economic criteria because the stock density was probably not affected by fishing. In the southern North Sea, because the stock was reduced by at least an order of magnitude, both drift netters and trawlermen went out of business.

In the present trawl fisheries for cod throughout the North Atlantic there are the obvious differences between nationalities, but no real ones in gear. Because the larger vessels range over a considerable area of the ocean there is a greater sense that the stocks are under international pressure rather than pressure by unfair gear. The trawlermen see the 'fish shops' develop in different parts of the North Atlantic, and on the crowded grounds the same ships see each other. Superficially, fishermen from one country will always shift the blame for heavy fishing to the fishermen from other countries, but in their hearts they know that on the open sea the pressure is international and therefore shared. The evolution of conscience is difficult to trace, but it must have occurred because no conservation can be achieved without the agreement of the fishermen.

The step in imagination from the relatively small catch of a single vessel had also to be made by the scientists to the stocks available in the sea. When Huxley said in 1883 that the seas were inexhaustible, he really meant that the ocean was very large indeed. The steam trawler with its otter trawl had only just been invented and the sailing smacks with beam trawls were confined to the North Sea although in summertime liner fishermen had been sailing into distant waters for centuries. It is difficult for us who cross the ocean in a few hours to conceive of the endless marine distance that faced our great grandfathers. Today we know that the catches taken by conventional methods are limited to about 100 million tons (Gulland 1972) and that the seas can be exhausted only too readily. Huxley could not have meant that the stocks were infinitely large, but only that they were large enough to stand the exploitation by the small Victorian fleets. When we transform his scale to ours, the great resources become diminished and the need for conservation presses.

The first step in conservation was given to me by a boat owner on the quay during the decline of the East Anglian herring fishery: 'Burn every b——'s boat but mine' However, this first step in realization is a considerable one. That it was needed appeared with the industrialization of fishing towards the end of the nineteenth century, in European waters and off the eastern seaboard of the United States. The International Council for the Exploration of the Sea was founded in 1902; the conservation of the fur seals on the Pribilov Islands started in 1911; and the first steps towards the conservation of the Pacific halibut were taken in the second decade of the century. The essential stages in the establishment of conservation in the North Atlantic were taken during the thirties, as were those needed for the conservation of whales throughout the world.

2. The small plaice problem in the International Council for the Exploration of the Sea

In the 1860s and 1870s the sailing smacks from British ports in the North Sea were formed into fleets under the command of an admiral. They stayed at sea for six weeks at a time and fish was taken to Billingsgate every day or so by carriers. The fish was transported from smack to carrier by small boats manned mainly by little boys, many of whom were drowned. In 1881 the first steam trawler was built and she fished independently of wind and tide; the beam trawl used by smacks was replaced by the otter trawl spread by two boards at the wing end rather than by the beam itself above the headline. By the end of the century more than 1800 steam trawlers had been built in Britain. Because they could work independently they abandoned fleeting. However, in the late 1930s some individual smacks were still working out of Lowestoft. Because of the earlier development of fleeting, the information available to the early trawlermen must have been enormous and they were able to exploit the stock of plaice in the southern North Sea with some efficiency. In particular they worked on the grounds close to the European coastline, which is the nursery ground for young plaice. There they caught fish that were undersized by present-day standards, but in summertime they discarded six times the quantity that they landed in their home ports. Garstang in the 1900s showed that with increasing numbers of steam trawlers the catch per trawler (or stock density) had declined by one-third in the last decade of the nineteenth century.

The industrialization of fishing in Britain started with the development of the railways. The history of Grimsby as a fishing port dates from 1849–51 when the railway reached the town. During the later nineteenth century, ice was imported from Norway and fish was carried into port on ice, and by train on ice to the developing industrial centres of the Midlands and to London. Fleeting from Barking on the Thames and from Grimsby was a further consequence of industrialization. The final step was the steam trawler which stayed at sea for ten days and brought back fish on ice that was good enough for the fish-and-chip trade. A similar pattern of industrialization occurred in the European countries, particularly in Germany. With industrialization came the need for conservation.

In 1899 a meeting was held by the Swedish Government in Christiania (Oslo) to consider the Exploration of the Sea. In 1902 the International Council for the Exploration of the Sea was established in Copenhagen. There were three branches to its activities, with committees on hydrography, fish migration, and overfishing; the Council undertook to make explorations in the North Sea with the research vessels of seven nations. But the variability of the trawl hauls inhibited the scientists and the name of the committee on 'overfishing' was changed to that of the 'small plaice problem in the southern North Sea'. As already noted, Petersen's solution was to conserve growth by transplanting young fish from poor feeding grounds to rich ones and Garstang concurred. In was left, however, to Heincke to face the small plaice problem properly. With Garstang's results he showed very clearly that the bigger plaice lived in deeper water and later this rule became known as Heincke's Law. If the little fish were to put on the weight they could, the shallow waters near the continental coasts should have been closed. The proposed closed area was the nursery for fish of one to three years of age where British trawlers had discarded them. Heincke proposed in 1913 that the area be closed *as an experiment.* After the First World War the Council never returned to Heincke's proposal. However, Faxa Bay in Iceland was closed experimentally after the Second World War. In the 1970s the idea of experimental conservation has returned to the international commissions.

A proposal was made in 1923 by a number of countries in the International Council through the Danish Foreign Office to close the small plaice grounds. In 1926 it was rejected by Britain because the British fishing industry, then the largest in the North Sea, was not convinced of its validity. In the Baltic in the late 1920s the stock densities of plaice declined sharply and later the catches also fell. Because the decreases were considerable the Baltic countries established a convention in 1933 by which the fishing for plaice was restricted. This was the first international convention in the Atlantic and it originated in the discussions of the small plaice problem in the International Council.

Petersen's solution to the small plaice problem in the first decade of this century was the classical statement of the problem of growth overfishing, that is, the little fish should be conserved to put on weight and so increase in value. He considered this to be a minor economic benefit that fishermen should be able to obtain for themselves merely by not fishing the small plaice. Today, fishermen from Lowestoft, from market considerations only, steam to deep water to find the largest plaice, bigger than those that are just retained by the present agreed mesh sizes. The real problem for Petersen was what we would now call recruitment overfishing. In the 1930s however, Graham and W. F. Thompson, reasoning with the logistic equation, suggested that recruitment should not be impaired within the normal range of fishing activity. Then the scientific problem was to mitigate the effects of growth overfishing.

3. The Pribilov seals

The conservation of the fur seals (and later the whales) has been included with that of the fish stocks because the biological principles are very similar and because the commission faces the same sort of international and intergovernmental problem. The life span of the fur seal is not so different from that of many fish, but their reproduction differs in that they produce one pup at a time and there is a territorial harem system which means that there is a surplus quantity of young bachelors, the holluschickie.

In the Bering Sea, north of the Aleutian Islands and hundreds of miles from Alaska, stand the Pribilov Islands. They are low and fog-bound, and in the early summer the fur seals come ashore to breed. First come the adult males, which mark out a territory and fight for it. Then come the young bachelors, who pass through the embattled territories on the beaches to the grassy areas behind. Lastly come the females, who on landing give birth to the pup that was conceived a year before on the same beaches. In the late summer, the seals leave the beaches and migrate across considerable distances in the Pacific before returning in the following year. The fullest account of the natural history and of the hunting is in Elliott, from which Kipling took his story of the White Seal.

The seals are exploited for their fur. Records of the catches go back for nearly two hundred years, for exploitation was started soon after the islands were discovered in 1786. The seals also bred on Commander Island, which is nearer to the coast of Kamchatka; apart from the two island groups, they breed nowhere else and for a life span of 12–20 years they return each year to the beach on which they were born. Because each bull has a harem which he defends with his life, there is always a stock of surplus males. These holluschickie have been exploited by the Aleut hunters for a long time.

After a period of very high catches in the early and middle nineteenth century, the islands were bought by the United States from the Russians and there was some control of the catches. However some seals were taken off-shore by pelagic sealers from Canada. A case was brought to the International Court in which the United States put forward the view that because the seals bred on the islands they were the property of the United States wherever they were caught outside the three-mile limit. The Canadian answer (through the British Government) was that the seals, like fish, outside the three-mile limit were free, to be taken by anybody. The Court found in favour of the Canadians, but cubsequently they agreed to stop catching the seals in the open sea and in exchange they took a proportion of the catch on the beaches and pelagic sealing ceased. The interesting point about this case was that it was the first statement of the proposition that a stock belongs to the state in whose waters it is found. The Pacific salmon that spawns in North American rivers is very vulnerable to exploitation in the Alaska gyral and so far the stock has not been fished in the open sea (apart from some wanderers in the western Pacific). Both salmon and seal stocks breed on United States territory, not even in territorial waters. The present case for the coastal state is only to extend territorial waters for fishing purposes and says nothing about the ownership of spawners.

In 1911 a treaty was signed between the United States, Russia, Canada, and Japan to establish the International Fur Seal Commission.

There were heavy catches in two seperate periods in the middle and end of the nineteenth century. After 1911 catches were restricted to a low level and in subsequent decades they were allowed to build up slowly. Japan left the Commission during the Second World War but returned when a reconstituted Commission was established in 1958. An analysis of the stock at different stages during its recovery showed that there was a dome-shaped curve of recruitment on parent stock and that the present catches are probably quite near the maximum sustainable yield. The reduction of recruitment at high stock was attributed to relative food shortage for the pups in the waters close to the islands. There is also a little evidence, however, that there is a greater mortality of the pups at high stock owing to a hookworm infection. The biology of the fur seal is very different from that of fish, but the form of population assessment is fundamentally the same. Indeed the explanations used to account for recruitment in fur seal and fish are identical.

The Fur Seal Commission was the first international one to be established and it is probably the most successful. The scientific problem was perhaps easier than those that usually face fisheries biologists, because the beaches accommodate the whole stock and they can be counted there. The number of cows per harem and the number of pups per cow could be estimated. Further the catch is counted in one place and is not sampled by a variety of different methods in different countries as so often happens with the fish stocks. However, although seals can be aged, ages are not determined in the routine and intensive way that is characteristic of some fish stocks. Still the most important factor is that the catch and the stock can be seen at the same time and any persistent overexploitation would have been very obvious. Another point is that the stock of surplus males is also distinct because they live apart from the harems on the beaches; man can replace the natural mortality of the males from the age of first capture to the age at which they take the harems over. Because so much of the population structure is displayed on the beaches, Elliott was able to make an estimate of the catch in 1884 that is not so very different from the maximum sustainable yield determined today with a stock–recruitment formulation.

4. The Pacific halibut

The halibut was well known to the Indians off British Columbia and later it was exploited by line fishermen of Scandinavian origin working out of Seattle and Vancouver. The fishery did not develop very much until the transamerican railways were opened. It happened that the stock of halibut in the north-west Atlantic had been depleted and, with the high demand, the catches of the Pacific halibut were increased and shipped east in cold storage. Peak catches were obtained at the beginning of the second decade of the century and subsequently they declined; stock density had fallen since the expanded fishery got under way.

There are two spawning grounds, one off the southern part of British Columbia and the other off the southern coast of Alaska; being so far apart, they are quite distinct. Fish tagged on the respective spawning grounds are recaptured within the area of liberation and do not travel far enough to enter the other; indeed the spread of tags off British Columbia is limited to a range of tens of miles, although off Alaska they may spread up to a few hundreds of miles. There is also a morphometric difference between the two groups. On this evidence they have been considered as separate stocks although the extent of interchange, if any, has not yet been estimated. However, there was only one fishery to start with and when the Halibut Commission was established it concerned itself with the fishery on the two stocks. In more recent times halibut have been found in the Bering Sea, and some of these that may belong to the Alaskan stock are caught there by trawl.

During the First World War, catches off British Columbia had fallen to a half and stock by a factor of three and discussions were held between

Canada and the United States on the state of the stocks. Later a treaty was signed between the two countries, and the International Halibut Commission was established in 1924. The first director was a well-known fisheries biologist, W. F. Thompson, who with the help of his staff advised the Commissioners on what action to take. This was the first of three species commissions to be established on the initiative of the United States; the other two are the salmon and tuna commissions in the Pacific. Not only is each commission limited in its activity to one species, but the director and his staff are considered to be international civil servants giving impartial advice to the commissions. Elsewhere, the commissions are advised by scientists in much the same way without conferring on them any quasi-international status.

Thompson and Bell showed a clear relationship between stock density (as catch per 'skate', or six hooks) and fishing effort (as number of skates). Hence a reduction in effort on this stock should increase stock density and also catch, provided that the reduction was small. Not only is such a small reduction desirable to retain fishermen in employment, but also with small consecutive steps in the regulation, the maximum catch can be attained. The same philosophy appeared later in the application of the logistic curve by Graham and Schaefer. Thompson restricted fishing effort by licensing the boats and closing the fishery for a season, starting in 1930. Subsequently in the British Columbian group, both stock density and catches rose steadily for two decades. The same thing happened in the group off Alaska, but the changes were not so definitely marked (Fig. 7).

The Pacific halibut was the first stock of fish to be conserved internationally; that of the Baltic plaice started a year or so later. The result of conservation was plain to see in that both the total catch and the stock density recovered by factors of 1·5 and 3·0 respectively; consequently, turnover in the industry should have increased and also the profit to individual fishermen. However, there were two economic disadvantages: first, the closed season was gradually extended and some boats spent only a relatively short time at sea, and, secondly, the licensing system perpetuated the old methods of fishing. There were further disbenefits in having to maintain the gear and to keep cold stores running longer than need be. The real issue raised here was that the fisheries biologists could well devize regulations to cause recovery of the stock, but that they might not see all the economic consequences of their actions.

One of the characteristics of the North American species commissions was that the science of the founding directors tended to be preserved. In

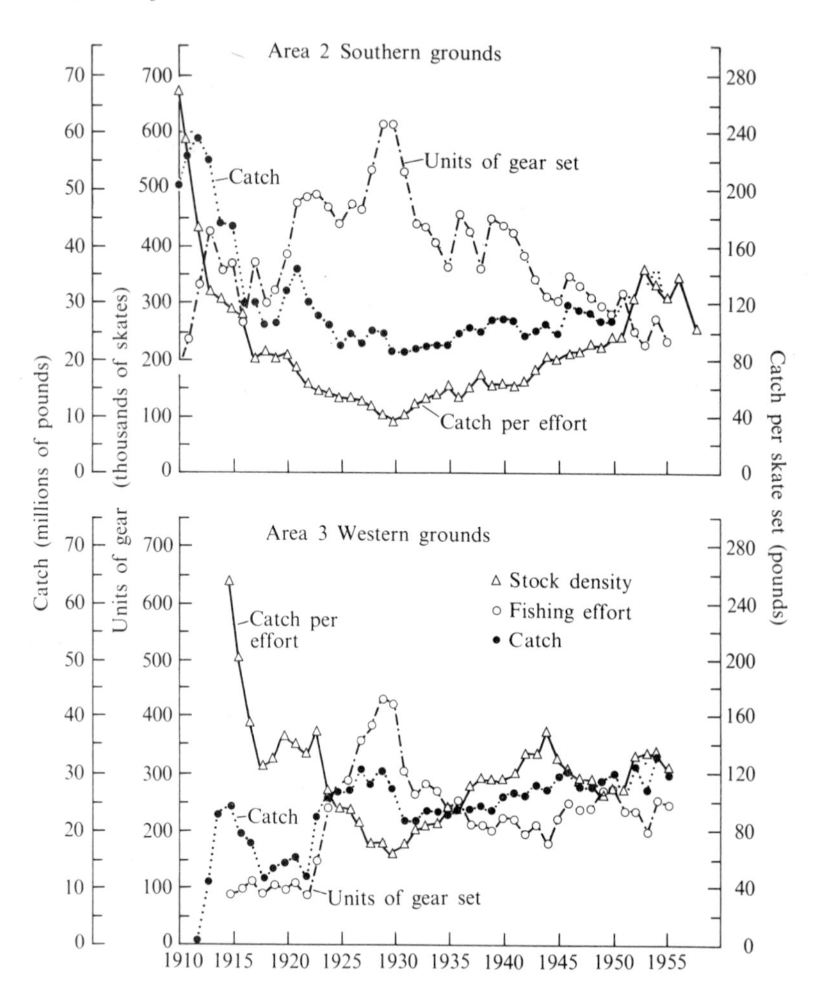

Fig. 7. The trends of stock density, fishing effort, and catches in Pacific halibut since the second decade of the twentieth century.

the Halibut Commission the methods that Thompson used in the twenties persisted into the late fifties and early sixties, when the Schaefer model might have been the logical extension of Thompson's philosophy. In 1917 Thompson had discovered how to age the halibut using the annual rings on the otolith. When he devized his method of stock analysis that depended on stock density and fishing effort, he concluded that age determination was not needed. Fortunately the otoliths were collected and were read some decades later; however his decision was preserved for too long.

Because the Halibut Commission was the first international one to conserve a fish stock its actions may have been analysed too far and with too great a degree of hindsight. It was Thompson's considerable achievement to define the stocks, to devise a model for control and to put it into operation. Later (Thompson 1950) he claimed to be lucky in that the two nations involved shared a common language, a common structure of law, a simple and single fish stock, and fishermen of common Scandinavian stock. In some respects, but not all, Thompson's procedure remains a model to be envied.

5. The International Council in the 1930s

By 1933 the Baltic plaice convention had become effective, but it was the only step taken by the International Council during its first 30 years of existence to counter the effects of heavy fishing. Heincke's proposal for an experimental conservation was never put into effect because of the outbreak of the First World War, and his successors in the early twenties were not able to convince the British fishing industry that conservation was needed. However, during the 1930s the foundations of the present structure of international commissions were laid. The scientific basis for this structure was that a trawl fishery could be regulated in general terms for all the species caught, which is a radically different approach from the species commissions of the west coast of North America. It developed from the solution of the problem of growth overfishing.

The scientific initiative for this development was British. Russell had published his statement of the overfishing problem, i.e. that the decrements in stock during a year due to death, including fishing, should be balanced by the increments in growth and recruitment. Graham (1935) put forward the model based on the logistic curve by which the greatest catches could be determined. However, as noticed earlier, the problem that faced the fisheries biologists was really the small plaice problem: how to let the little fish escape to put on more weight before they were finally caught–how to conserve growth; this was effectively how Graham expressed his model.

Some trawls had been designed specifically to let little fish escape, for example, the 'Trouser' trawl of Russell and Edser, but the general solution was to increase the mesh sizes in the cod end, the bag at the end of the trawl where the captured fish collect. Davis devised a cover to a large meshed trawl, so that the fish escaping from the trawl were retained in it. Using this method he was able to show that with an increased mesh size only the larger fish were retained by the trawl. Hence the solution to the problem of growth overfishing was to determine minimum mesh sizes. In 1933, the British introduced a minimum mesh size in addition to the minimum landed sizes (for different species) that were already in use. The fishermen were not all convinced that the control of mesh size was reasonable, because they thought that the meshes closed up as the trawl was drawn through the water. This point was not answered to their satisfaction until the open meshes were shown in an underwater film in 1950. However the distributions of fish retained by the trawl and those collected by the cover showed that escape was taking place and that the selectivity of the trawl could be measured.

The problems of the growth of fishes, of undersized fish, or mesh selection, and of overfishing in general were discussed by the Council during this period. In 1937, a Convention was held in London on the protection of undersized fish and it was signed by ten countries all of which had participated in the work of the Council. It was not ratified and was effectively replaced after the Second World War by the Overfishing Convention of 1946. During the 1930s, the International Council decided that the science was far enough advanced to persuade governments to start the negotiations that eventually controlled the effect of fishing.

Because the trawl catches all the fish on the sea bed in a catholic way, the principle of mesh regulation could be applied to all demersal fish. Consequently species commissions were not established in the north-east Atlantic. The International Council was already in existence and the scientists were already accustomed to working with each other; there was no need to establish an international body with a director, as for example in the Halibut Commission. The difference between the species Commissions and the later Atlantic Commissions arose in the way in which the scientific problems were solved. Thompson restricted fishing effort by means of the dependence of stock density on effort in a single stock and he used a closed season; Graham, thinking beyond the small plaice problem, devised mesh regulation to conserve growth in all demersal stocks and not put fishermen out of work.

During the 1930s the International Council was also concerned with whaling. The important paper by Hjort *et al.* on the application of the logistic curve to whale and fish populations, was published in the Norwegian journal of whale research *Hvalradets Skrifter.* The Norwegian government had prepared draft Bills on whale conservation and they were discussed in the International Council. Whaling in the North Atlantic was an ancient and moribund industry, but it throve in the antarctic from the middle 1920s onward. Considerable concern was expressed on the state of the Antarctic whale stocks and in the Council it was proposed that an International Whaling Convention should meet, which it did in 1937. The Second World War prevented any ratification but in 1946 a Whaling Convention met in Washington D.C. and the International Whaling Commission was subsequently established in 1950.

The decade was a period of fulfilment in the International Council. After an initial time of dismay the scientists came to understand the variability of catches with the statistics of small samples. It became possible to manage the market samples in such a way that the population biologists could detect the trends in stock density and at the same time point to the loss of older and bigger fish with the increased pressure of fishing. This was Graham's most telling point in his argument to conserve growth (Graham 1973).

The International Council embraced the thesis put forward by Graham and laid the foundations for the two Atlantic Commissions, the Northeast Atlantic Fisheries Commission, and the International Commission for Northwest Atlantic Fisheries. The earlier Baltic convention played a part here on an earlier scientific case, for its intergovernmental success encouraged others to follow its lead. The same optimistic climate saw the start of the debate that led to the establishment of the International Whaling Commission. Thus there are four international commissions that owe their origin to the work executed in the International Council during the 1930s.

6. The establishment of international conservation

The conservation of whales and seals has been included in this account of the management of fishery resources because the sciences are similar and the problems confronting international management and the population experts on seals apply their skills to fish stocks. It was no accident that the Whaling Commission was the only international forum for the discussion of the science of marine resources. On the other hand, the Fur Seal Commission originated long before any of the fisheries commissions and remains an example to them.

The fisheries commissions owed their origin to the industrialization of fishing with the advent of steam trawlers in the North Sea and the introduction of cold stores in Seattle and Vancouver. Laboratories to investigate the biology of fishes were established in many places during the late nineteenth century, but the biologists to investigate particular fisheries were not engaged till later. Indeed the Lowestoft laboratory was established to work up the material from the trawl surveys conducted by the International Council for the Exploration of the Sea in the first decade of the century. The Council itself was founded in 1902 in response to concern felt by fishermen and others on the state of the stocks in the North Sea. In general, fisheries biologists learned to investigate fisheries on stocks that had already been well exploited.

The success of the Fur Seal Commission probably originated in the intuitive understanding that the commissioners had of population dynamics. They were American, Canadian, Japanese, and Russian and the disparity in law and in language was as great as could be. Yet the agreement was established in 1911 long before there was any habit of international concordance. Elliott had formulated the problem some 30 years before and the commissioners were almost certainly aware of it. They may have visited the Pribilov Islands and Commander Island and seen that a proportion of the male population could be taken. With the long record of catches before them it might have been obvious that to obtain sustained catches that proportion should not be excessive.

Thompson attributed the success of the Halibut Commission to the common language and law in the United States and Canada and to the community of fishermen of Scandinavian origin. These factors helped, but a more important one may have been that with his simple model he could explain what had happened to the stock in a clear and convincing manner. Intuitively it is obvious that as fishing increases the stock density must decline and that if greater catches are needed the stock density must be allowed to recover. Just as a great advocate expresses the intricacies of the law in a simple and convincing way to convince a jury, so must the fisheries scientist convince his commissioners in the simple way that appeals to their intuition.

In the International Council for the Exploration of the Sea, a poor start was made as far as international regulations was concerned. Had they studied catches on the markets rather than the tenuous results of a groundfish survey, their disappointment with the variability of stock density might have been tempered. However, during the 1930s the scientific foundation was laid for both the later Atlantic Commissions (including the Baltic convention for plaice) and for the later Whaling Commission. This scientific base is found in the papers of Russell, Hijort *et al.* and Graham (1935) and the model used was that of the logistic equation. But, historically, the Council remained dominated by the 'small plaice' problem posed in the first decade of the century. Petersen had shown that if fishermen allowed the little fish to escape, they put on weight and subsequently the catches would be of greater value. The converse and convincing step was made by Graham in showing how an excessive fishing rate deprived the stock of older fish and how the 'small plaice' problem was generated. His solution led to the conservation of growth and to the solution of the problem of growth overfishing.

Equally important was the demonstration that the 'small plaice' problem could be solved effectively by merely increasing the mesh size of the cod ends in the trawls. Hence the fishing rate was reduced without endangering the employment of the fishermen and the mean size of fish in the catches was increased. All these events occurred later, but the solution was visible to the scientists working in the International Council and to the fisheries administrators who comprised the Overfishing Convention of 1946. The essential piece was the very simple idea that if small fish escaped from the cod end to put on more weight their period of vulnerability to fishing was shortened. Subsequently Beverton and Holt (1957) showed in a rigorous manner how stock density, total catch, and mean size would all rise with increasing mesh size to limiting values. In the 1950s and 1960s there was much argument about the details of mesh regulation, but in the 1930s the important point was the clear and simple idea that carried conviction because it was this that led the fisheries administrators to action.

In the early steps towards conservation on an international scale the simple scientific models played a considerable part. There were four international bodies established before the Second World War, the International Council for the Exploration of the Sea, the Fur Seal Commission, the Pacific Halibut Commission, and the Baltic Convention. The International Council is not a regulatory body, but it was responsible for establishing the Baltic Convention through the Danish Foreign Office and it played a great part in the movement towards the Overfishing Convention in 1946: its meeting in May 1939 in Berlin before the Second World War was on the 'Overfishing' problem. In the Fur Seal Commission the model was simply the ratio of surplus males to total stock. In the Pacific Halibut Commission and the Baltic Convention it was the inverse dependence of stock density on fishing effort. In the International Council the simple model stated that if fish escaped to put on more weight they decreased their chance of death. There is of course a concealing art in simplification, but to carry conviction models should be simple. It must not be denied that much of the work that contributed to the models was difficult; for example the analysis of tagging experiments is complicated and sometimes unrewarding. In the early steps towards conservation the successful models were simple to understand.

5

The international structure of management

1. Introduction

In the years after the second world war an echo was heard of Huxley's inexhaustible sea (Huxley, 1883). From European or North American cities men looked from the exploited stocks near their shores to the unexploited ones in waters distant from them. There is a chart largely drawn by Graham in 1949made at the UNSCURR meeting in Rome. Some stocks in the North Atlantic are shown as being fully exploited but in the rest of the world there are shown great areas of unexploited stocks of sardines and hake. Most are in the four major upwelling areas, in the California current, the Peru current, the Canary current, the Benguela current, and the rest were in the Indian Ocean or in Indonesia. Another chart drawn by Gulland (1972) only twenty years later shows that many of these resources, but not all, are now well exploited. During the late fifties and early sixties it was often thought that man would obtain a large increment of protein from the sea, but Gulland has shown that the limiting catches of about 100 million tons per year might be reached in a decade or so. In the 1930s the shelf seas off north-west Europe or off the north-west coast of North America showed signs of exhaustion; today it is the world ocean that might be limited in yield.

The development of these resources has been obtained by the growth of fishing fleets. Before the Second World War the traditional ways of preserving fish to bring it ashore were either to put it on ice (and call it fresh) or to dry it in the sun and wind. During the 1950s and 1960s freezers were built into stern trawlers. They were necessarily larger vessels, with more crew and greater range and they could stay at sea for two or even three months. In Europe there are still fleets of side trawlers, or side winders, that pack their catches away on ice, but they are gradually being replaced by stern trawlers that freeze the fish. Such large stern trawler fleets are also operated by the Japanese, the Russians, and a number of Eastern European countries. Together they form the basis of the mobile, long-ranging, and international fleets that travel all over the world ocean.

Another development originated in the broiler chicken industry. The chickens are fed largely on fish meal because it is the cheapest protein available. Vegetable protein can also be cheap but lacks two essential amino acids, lysine and methionine, that occur naturally in fish meal. Most of the meal is made from herring or sardine-like fishes that are caught very efficiently by purse seine. About 15–20 million tons of fish each year are converted to fish meal, capelin from the north-east Atlantic, menhaden from the eastern seaboard of the United States, pilchards from South and South West Africa, and anchoveta from Peru. Because such fish are cheap, because they appear to be superabundant, it has been thought that the stocks were not very vulnerable to fishing. In fact, as shown earlier, the herring and sardine are prone to the most disastrous form of overfishing, recruitment overfishing. The herring fishery in the North-east Atlantic has practically disappeared, taking with it for ever all those diverse ancillary trades that depend on it. They supplied herring for people to eat, but more profit was gained if chicken ate them first. The pilchard fishery off South Africa and off South West Africa has become much reduced (*FAO year book* 1971) as has that for menhaden off the United States. In 1972, the recruitment to the largest fishery in the world failed, the Peruvian anchoveta of which about 10 million tons were landed every year. Such stocks are in urgent need of conservation even if fish meal is their only outlet. Those off Peru and off South West Africa are under control by catch quota; in 1973 a stock–recruitment relationship for the Peruvian anchoveta was established. Had the fisheries developed more slowly, it might have been established before the recruitment failure.

In Japan and Russia decisions were made in the early 1920s to increase fish catches to provide more protein for the people. Exploratory voyages were undertaken all over the world to be followed by commercial fleets. Russian trawlers work on most coasts of the North and South Atlantic and on many Pacific coasts including those of Canada and the United States. They catch cod on the traditional North Atlantic grounds, hake off South Africa, North Africa, the east and west coasts of North America and Alaska pollack from the Bering Sea; the geographical range of their fleets on one important family of fishes, the cod-like ones, indicates the scope of their activities. The Japanese have considerable trawler fleets also, but their tuna fleets travel the greatest distances. The tuna are caught in all parts of the subtropical Atlantic and Indian Oceans and in most parts of the Pacific. They are frozen and exported to an avid market in the United States. When the exploitation of tuna was first considered it was thought to be an enormous resource because the ocean is after all very large; however, the Japanese, the Koreans, and the Taiwanese have reached the maximum for certain species in the Atlantic and the Indian Oceans and perhaps also in the Pacific. Between them the Japanese and the Russians land about 12 million tons each year, about a fifth of the world catch. Their explorations have been followed by other nations: Americans catch tuna off Angola, Spaniards catch shrimp off Mozambique, the French caught crayfish off Brazil and so on.

The great expansion of fleets of large and mobile vessels and the dominance of the frozen fish and fish meal outlets has exploited most of the world fish resources. However there are still quite large resources still to be discovered and caught in the Indian Ocean and the Indonesian area. At the present rate of increase that limit of about 100 million tons will be reached within a decade. Some of these resources are already conserved fairly well, but the conservation must be extended to all of them and quickly.

2. The Atlantic Commissions

As a result of the Overfishing Convention of 1946, the two Atlantic Commissions were established, ICNAF (International Commission for Northwest Atlantic Fisheries) in 1949 and NEAFC (North East Atlantic Fisheries Commission) in 1954 (as the Permanent Commission and then in 1964 as NEAFC). In both, the initial steps were made in mesh regulation and the scientific model was the yield per recruit one of Beverton and Holt. By the early sixties most of the demersal stocks were controlled by minimum landed sizes in the ports and minimum mesh sizes at sea. There was much discussion on the technology of mesh regulation but in the end the larger meshes were accepted.

However conservation by mesh regulation is least conservation because nearly all trawl fisheries are mixed. Then the mesh of the trawls is that for the smaller fish of interest to the fishermen. The 80 mm mesh in use in the North Sea yields good conservation for haddock, but for larger fish like cod bigger catches would be obtained with much wider meshes. Good conservation could be obtained by mesh regulation in an unmixed fishery, but in the North Atlantic trawl fisheries there are species other than cod. On the northern grounds in the Atlantic the best mesh size for haddock is a little less than that for cod. There is a sense in which mesh regulation, like the minimum landed size, should prevent failure in conservation.

Mesh regulation by itself does not prevent the uncontrolled entry of new fishing vessels each using the right mesh size. Then the yield might be reduced to low levels by recruitment failure. For example, the catches of the arctonorwegian cod stock rose to 800 000 tons in the early sixties and the stock may well have been reduced far enough to cause a sharp decline in recruitment (Garrod 1967). However this event occurred before the North Atlantic scientists realized that fishing might adversely affect recruitment.

In 1967, a new technique in fisheries science was introduced as noted earlier, cohort analysis by which catches at each age in a cohort were used to estimate fishing mortality in each year. Annual estimates of stock emerge by age and hence estimates of recruitment. The whole population structure of catch and stock by age is then revealed in absolute numbers which can be readily converted to weight. Almost as a consequence, catch quotas were introduced for certain stocks in the north-west Atlantic in 1970. It is a most important advance because with suitable annual adjustments, the best yields should be obtainable for each stock and each species. Management by catch quota has only just started, and it will take a little time before the best results can be obtained. Earlier it was noticed that conservation had sometimes been inhibited by scientific failure, but this is a considerable success. In the same period, a notable institutional step was taken in that international enforcement was introduced throughout the North Atlantic (1969 in ICNAF; 1971 in NEAFC). Naval vessels of any participating nation could stop any fishing vessel of the treaty nations on the high seas and inspect their gear; any case of infraction of a regulation is brought in the fishermen's own country. This surrender of a little sovereignty in the interests of conservation shows only too clearly how seriously it is taken by the nations in the North Atlantic.

Despite initial misgiving about the details of mesh regulation (there were never any about the principle) the present system of minimum landed size and minimum mesh size, with the addition of catch quotas and a system of international enforcement, seems to hold considerable promise. About 20 years after the establishment of the two North Atlantic commissions there is now a confidence that they will proceed from a least or moderate form of conservation to something near a very good one. The credit is due jointly to the scientists and commissioners who work closely together. Where there has been failure, as for example, in the temporary decline of recruitment to the arctonorwegian cod stock, it is because the problem of stock and recruitment caught the scientists unprepared.

However there has been a considerable failure in the north-east Atlantic with the loss of the herring fisheries. In the North Sea there were three stocks of autumn spawning herring all of which were reduced between 1955 and 1965. In the Norwegian Sea there was the large stock of atlantoscandian herring, the fishery for which ceased to exist in 1968. Some stocks remain, for example, to the west of the British Isles and in the Celtic Sea. The first stock to disappear was the Downs stock that spawned in the eastern English Channel and the southern North Sea; however, with much relaxed fishing effort, numbers of larvae started to increase again in the late 1960s. Hence there is some evidence that the stock suffered recruitment overfishing. This was the conclusion put forward earlier and it is likely that the atlantoscandian stock suffered the same fate.

Herring do not grow very much during their adult lives and the curve of yield-per-recruit is asymptotic to it at a fairly high level of fishing mortality. From this simple picture, it might be concluded that herring could be fished heavily with no fear of failure. However, it is likely that the stock–recruitment curve of herring is such that recruitment can be reduced at relatively low levels of fishing. But such a relationship was not formulated until after the herring fisheries had declined and some had indeed disappeared.

Although Beverton and Holt (1957) recognized the stock and recruitment problem and advanced it, they were also very much aware of the high variability of recruitment. The application of the yield-per-recruit model implied that it should be used until recruitment were shown to fail. From our present knowledge of the stock and recruitment problem, such a criterion would be used much too late. For cod-like fishes, the maximum yield would have been well exceeded and for herring-like fishes, the point of recruitment failure might have been reached. In the Atlantic commissions, the success of practice in conservation probably originated in the yield-per-recruit model and as a consequence the demersal stocks are quite well protected with the one exception of the arctonorwegian cod stock. With the introduction of cohort analysis they will be very well protected in the future. The failure in conservation in the Atlantic arose from the implication of the yield-per-recruit model that it was safe until recruitment failed. The cod-like fishes are much more resistant to recruitment failure than the herring-like ones and it is no accident that by and large they have been conserved successfully whereas many of the herring fisheries have been lost. Thus failure in the commission is linked to scientific failure. But as indicated above success is linked to scientific success. This link between science and management is a most important one and will be pursued later.

3. The Pacific Commissions

The International North Pacific Fisheries Commission was established in 1950. An important part of its practice is the abstention principle, which states that a potential entrant to a fishery should abstain from fishing if the stock is already fully exploited; that is, at the maximum sustainable yield. For example, the North American salmon and halibut stocks were shown to be fully exploited, so the Japanese agreed not to catch them. The southern Alaskan herring was considered underexploited and potential entrants to the fishery, if any, would have been permitted. The abstention principle illustrates a point made by economists, that outside territorial waters fish are common property and are not rentable. Because they are not rentable, market forces cannot operate and it is therefore necessary to control entry. To consider the point on biological grounds only, we have already noticed that the arctonorwegian cod stock suffered an increment of effort in the early 1960s that may have damaged recruitment subsequently. In any commission the possibility of uncontrolled entry is feared considerably. The abstention principle was not devised for economic reasons, but to protect the fishermen of North America from the overexploitation of stocks in which they were interested.

The Halibut Commission has already been described. There are two other species commissions in the North Pacific, the International Pacific Salmon Commission was established in 1937 under the direction of W. F. Thompson. The Tuna Commission was formed in 1949 and M. B. Schaefer was its first director. The salmon and halibut commissions work both independently and within the North Pacific Commission. The tuna commission is quite independent of the INPFC. All three have a director and a staff paid by the commission and therefore are considered to work independently of nationality.

When the Transcanadian Railway was built, it passed through a defile on the Fraser river called Hell's Gate, and there was a landslide in 1913 that prevented the sockeye salmon from swimming easily up the river. One of Thompson's first acts as director was to demonstrate the destructive effect of the Hell's Gate landslide on the stock and to mitigate its effect by building a fish way for the salmon. Since the end of the Second World War when it was built the stock of sockeye in the Fraser river has increased considerably. The fifty stocklets or so in the Fraser river system can be distinguished by the number of circuli (or minor rings) on the scales. Before they ascend the river system, the salmon gather and are caught between Vancouver Island and the mainland, and many of the different groups can be identified in the catches. They enter the fishery in an order corresponding to the distance that they have to travel and they move up the river at speeds also related to that distance. This discovery is a most remarkable one to be added to the successful restoration of the stock.

The main international task of the Salmon Commission was to discover where the immature salmon live in the open Pacific ocean and to separate the Asian and North American stocks there. The Japanese fished for them in the western Pacific since the 1930s, but after 1950 they agreed to abstain from fishing them in the Alaska gyral in the eastern Pacific. In an early decision of the INPFC they agreed to abstain from fishing salmon east of 175° W, but of course were free to continue fishing in the western ocean. In subsequent years there was considerable argument on the degree of overlap between the two groups. Evidence from the distribution of tagged fish, from the distribution of an Asian and a North American parasite and from the statistical study of morphometric characters showed that the early decision was probably right and that the degree of overlap between the stocks extended across some ten degrees or more of longitude. Cushing (1972) gives a short account of this work. Later the abstention on this point was not revoked.

In the Bristol Bay stocks of salmon that enter Alaskan rivers from the Bering Sea an important management technique has been developed. It is not really part of the Salmon Commission's international work, but the method might be extended to it (and elsewhere) at any time. A forecast can be made of the stock, or run, and an estimate of the catch is made in time, taking into account various environmental and economic factors. Normally, the management of such fisheries is limited to generalities such as closed seasons, mesh sizes, and catch quotas. However, the application of systems analysis to the somewhat specialized problem of the Bristol Bay salmon is more or less successful quantitatively and economically. It is possible that such techniques will be eventually applied elsewhere in fisheries management, particularly following the development of the catch quota system in the North Atlantic.

When the stocks of anchovies in the Central American bays of the Pacific failed, the Tuna Commission was established to investigate their disappearance because the tuna boats used the anchovies as bait. However after a very few years the boats dispensed with the bait by turning to purse seining. The Commission, faced with an expanding fishery needed an estimate of the maximum sustainable yield of the yellowfin tuna. The director of the Commission, Dr. Schaefer, extended Graham's use of the logistic curve and devised what is now known as the Schaefer model. The yellowfin spawned in the eastern Pacific off the Revilla Gigedos Islands; there are certain morphometric differences between them and those caught further west and tagged fish were nearly always recovered in the eastern tropical Pacific. With the stock problem provisionally solved, Schaefer used his model to calculate a maximum sustainable yield to which he attached confidence limits. Because the tuna cannot be easily aged, the logistic curve provides a very useful method. In recent years the catches have expanded beyond the upper confidence limit to a new maximum which suggests first that the stock problem was not in fact solved, and secondly that a slow expansion can be allowed.

Conservation in the North Pacific is in three parts (*a*) the North American species commissions (*b*) the International North Pacific Commission (*c*) Russo–Japanese commissions for parts of the north west Pacific. The species commissions take care of three species in some areas; the halibut in the Alaska gyral is conserved, but not if the fish migrate through the Aleutian into the Bering Sea; North American salmon are probably well managed, but the Asian salmon catches have been declining for years; the yellowfin tuna in the eastern tropical Pacific is being exploited carefully but there is no control yet for all the tuna species elsewhere in the Pacific. The North Pacific Commission itself has limited the catches of king crab in the Bering Sea, in addition to its activities through the species commissions. Stocks of Alaska pollack, flatfish (or *Limanda aspera*), and black cod are exploited without let in the Bering Sea. There is no control in the waters round Japan, the east China Sea, or the Yellow Sea, where stocks have long been overfished. There are bilateral arrangements between the U.S.S.R. and the United States that allow the Russians to fish off the American coasts. There is no overall system of international control in the Pacific, like that in the Atlantic. The INPFC could be the body to take the control, to introduce a system of international enforcement and perhaps to establish methods of conservation, such as catch quotas.

The reason for the lack of generality in the International North Pacific Commission is compound. The prime reason is that North American fishermen have not worked very far from shore until quite recently. The species commissions were established to protect three rather small groups of fishermen on the west coast of North America. The abstention principle arose from the need to protect a coastal fishery for salmon. The bilateral treaties negotiated between the U.S. and the U.S.S.R. apply a temporary and limited conservation to stocks only lightly exploited by the North Americans. The second reason for the limited outlook of the INPFC is that until very recently the Japanese have thought it was much more important to exploit a stock than to conserve it. The third reason is that many countries in the North Pacific are not represented. The stocks fished by North American fishermen are well managed, but the rest, or the large majority, are not managed at all.

4. The international Whaling Commission

The International Whaling Commission was established in 1950. Research in the participating countries (Japan, Netherlands, Norway, U.K., U.S., U.S.S.R., South Africa) was co-ordinated in the Commission and the stocks of five species were controlled by a catch quota of Blue Whale units, weighted by the sizes of the different species. During the 1950s catches of blue whales declined dramatically despite the continuous application of slowly-reduced catch quotas. Later the catches of fin whales fell in a similar way. The collapse of the blue whale stock has become a *cause célèbre* for the political ecologists.

In 1956, Ottestad diagnosed the cause of the trouble by saying that the recruitment was declining under the pressure of fishing. Such estimates required that the whales could be aged and at that time there were three independent methods of age determination. The Dutchman, Schlijper, proclaimed that they were unreliable and Ottestad's diagnosis was not accepted. Between 1961 and 1964, a committee of fisheries biologists was invited to tackle the problem. They appeared to evade the question of age determination by applying the Schaefer model directly (Anon 1964): it was shown that the maximum sustainable yield had been reached in the 1930s and that since then the stock had declined with the increased pressure of fishing. The yields at low stock and high fishing were well fitted by an estimate of recruitment, less mortality (Fig. 8). Ottestad's diagnosis was right in that recruitment was reduced by fishing, but Schlijper no longer disputed the point, although the same age determinations had been used to estimate recruitment.

Although the scientific solution appeared in 1964, the catch quotas were not reduced to a level low enough to prevent further decline until 1968. The reason for the delay between solution and action was that the Japanese entered the whale fishery late having bought factories and

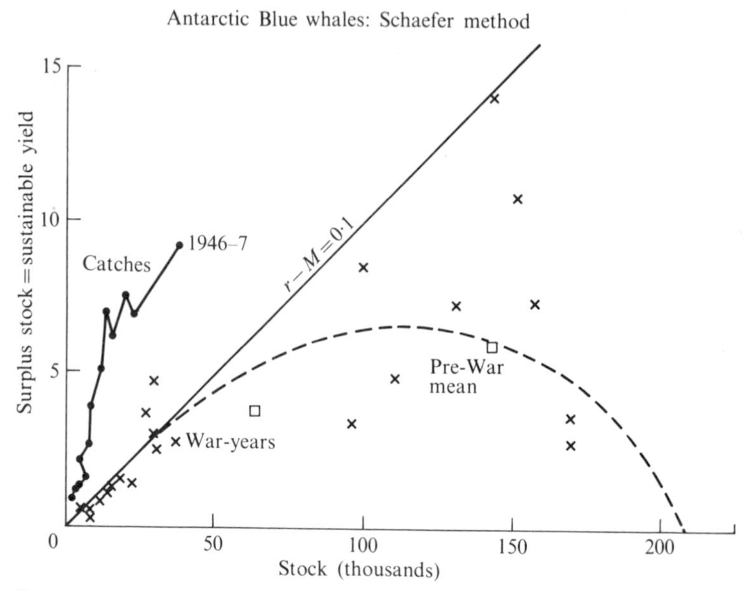

Fig. 8. The application of the Schaefer model to the blue whale stock; the straight line fitted through the origin represents recruitment less mortality (Anon 1964).

catchers from the Netherlands, Norway, and the U.K. when these countries stopped fishing; each delay went some distance towards the amortization of investment by the Japanese. However, during this interim, the Commission encouraged the whalers to divert their activities from the blue and fin whales to the humpback and the sei. The four fisheries biologists showed that the sei and humpback could stand further exploitation. By 1968, when the capture of blue whales was forbidden by the Commission, the stock had been reduced to about one thousand animals. The confidence limits to any such estimate were high and many people thought that perhaps the blue whale stock had been extinguished. No fisherman would waste his money in pursuit of the scattered remnants of a stock of fish. But it might just have been profitable for the whalers to kill the last blue whale.

Fortunately the Commission prevented this disaster. In March 1973 there were indications that the numbers of blue whales were increasing. Further, Gulland has pointed out that the catch quota imposed by the Commission delayed the decline in catches considerably; he estimated a saving of £60 000 000 since the inauguration of the Commission. To put it another way, some conservation is infinitely better than no conservation because without it there is always the threat of extinguishing the fishery, if not the stock. Fisheries such as the Californian sardine, the Japanese sardine, and the Norwegian herring vanished through recruitment failure without any conservation; each failure was attributed to rather unspecified natural events. Had fishing been reduced after the initial recruitment failure, we might have known whether the cause was natural or man-made. When the recruitment to the British Columbian herring failed, fishing was restricted as an experimental and conservative measure. Of course this procedure does not distinguish decisively between natural and man-made causes, for one could imagine an interaction between the natural effects and the effects of fishing. To put it more generally, if recruitment fails when fishing effort is high it is prudent to assume recruitment overfishing and then the best chance is taken to conserve the stock.

Blue whales produce one calf at a time. Therefore the dependence of recruitment on parent stock is easier to establish than in a fish stock in which recruitment is highly variable. The failure of the blue whale stock was a classical example of recruitment overfishing and the diagnosis was confirmed as the stock began to recover. Hence Ottestad's original solution to the problem was in principle right. But because it was vulnerable to Schlijper's criticisms, the Commission could not proceed. It only succeeded when they were overcome and the state of the stocks was neatly portrayed with the Schaefer model. Management was inhibited when the science faltered and could only act when the scientific solution was simple enough to carry conviction. It might be protested that the Commissioners put their countries' needs first and the care of the stocks last. Each country joins a Commission freely and is always free to leave. If there is scientific confusion or a complex case, the Commissioners tend not to agree and put the needs of their countries before that of the shared stock. But it is scientific failure and not an institutional one.

5. The need for management and the new commissions

The potential yield of conventional fishes from the ocean is probably about 100 million tons each year and the actual yield may be less than that for economic reasons. It might become possible to harvest antarctic krill, the shrimplike whale food and at a further stage of technological development to catch the small dispersed fishes of the deep ocean. But the catches of ordinary fishes with present gears are limited and at the present rate of increase the maximum should be reached in a decade or so (Gulland 1972) as noticed above. Therefore there is a pressing need for a full management structure in each of the oceans. There is such a structure in the Atlantic and the basis for one in the Pacific. Each set of commissions, including those for the whales and seals, has had its setbacks, as in any human institution, but all are by and large successful. That is, the step from no conservation at all to some conservation has been taken; if the best conservation has not yet been reached everywhere, most stocks are secure and for many, considerable progress is being made. Outside the commission structure stocks remain vulnerable and of course there remain some stocks in danger under the care of the commissions.

In the years just after the Second World War, FAO (Food and Agriculture Organization of the United Nations) established some advisory commissions, for example, the Indo-Pacific Fisheries Council and the General Fisheries Council for the Mediterranean. They are international scientific bodies at which scientists in a given region meet and discuss their problems. In purpose they resemble the International Council for the Exploration of the Sea and in their conferences research is discussed and to some extent coordinated as in the International Council. The latter body in the first half of its existence could initiate action through the Danish Foreign Office and later through NEAFC. In a similar way the advisory councils could initiate action through FAO although none was taken.

In the late 1960s and subsequently, FAO established a number of international commissions by inviting nations to take part, and by providing secretarial assistance and scientific advice. Then the participating nations agreed to establish the commission and endowed it with powers to regulate the fisheries under its control. The area between Dakar and the Congo is supervised by the new West Central African Fisheries Commission. It is of considerable importance because there are three upwelling areas in the region, which provide catches of *Sardinellas* and which are already exploited by vessels from outside the region. The Commission also embraces the eastern tropical Atlantic, a rich area of productive currents. Offshore there is the new Atlantic Tuna Commission whose aegis extends across all the subtropical Atlantic; the participating countries are Japan, Brazil, Korea, Canada, France, Ghana, Morocco, South Africa, Spain, Portugal, U.S.A.; the maximum yield of some species (but not all) has probably been reached already. To the south of the Congo there is the Southest Atlantic Fisheries Commission whose orbit extends round the coast of Southern Africa to

Moçambique. There is a large stock of hake in the Benguela current off South Africa and South West Africa and on the Agulhas Bank that is exploited by an international fleet of trawlers from South Africe, Japan, Spain, and the U.S.S.R., amongst others; offshore vessels also work on the pilchard stocks off South West Africa, which are very large. One of the newest commissions, and potentially one of the most important, is the Indian Ocean Commission which will be responsible for the fisheries present and developing in the whole of the Indian Ocean. It is in this area that much of the remaining development of world fisheries will be made.

There are other commissions that are proposed or in train of establishment, but there remain areas and stocks without supervision. There is no international control of the tuna in large areas of the North and South Pacific. Although there are advisory councils in the Mediterranean no action has yet been taken. There is no commission that might supervise the antarctic resources of krill or of fishes that might be exploited there. There are extensive areas off Indonesia, Australia, New Zealand, China, and Malaysia where there is no international supervision of the fish stocks. Perhaps commissions are not needed everywhere yet and they will probably appear as the need arises.

The largest fishery in the world is that for the Peruvian anchoveta, that yielded 10 million tons per year until 1972. It is exploited only by the Peruvians and is managed by them with scientific assistance from FAO. The stock is exploited and managed by the coastal state and at the present time there is considerable argument about the rights of coastal states to the fish stocks that lie in the seas adjoining their coastlines. The traditional territorial waters extended to three miles and after the Law of the Sea Conference in 1958 many fishing limits were extended to twelve miles from headland to headland (six miles territorial and six miles for exclusive fishing rights). The subjects will be disputed again at the second Law of the Sea Conference in 1974 and 1975. The opinion for the coastal state point of view arises from two sources. First, salmon that spawn or seals that breed on United States territory might be said to belong to the United States wherever they are caught. This viewpoint is not put forward by the United States government, but it appeared in the Fur Seal case of 1893 and may have influenced the development of the abstention principle in 1950. Generalizing, a stock that lives in the sea adjoining a coastal state is considered to belong to it. The second origin of the coastal state point of view is the sight of foreign fishing fleets offshore.

The only question considered here is how the coastal state attitude affects conservation. The most important point of all is that where failure has occurred in the commissions, it is usually a scientific failure of one sort or other. Very often the commissioners are blamed for taking a political view; this is their job and sometimes the failure in science is not sufficiently apparent. The same difficulty arises in coastal state management: when the recruitment to the Peruvian anchoveta stock failed nobody knew whether the cause was man made or natural or a mixture of both. Despite the fact that the Peruvians took a prudent course in restricting the fishery, the scientific failure is patent. The real problem is to make the science as good as possible as soon as possible irrespective of the institution of management.

If there is a conflict between commission and coastal state, it is almost a geographical one. Fish stocks inhabit natural oceanic regions, the Peru current, the Alaska gyral, the Norwegian sea, and the largest of these is the North Pacific subtropical gyre. A small area like the North Sea with seven or ten countries taking part in the fisheries and the stocks swirling in a single system can only be managed by a commission. The long coast of Peru with a stock of anchoveta and a stock of hake, with no international participation at sea, is managed by the coastal state. Between the two extremes, there are a number of compromises; for example, the bilateral agreements between the U.S. and U.S.S.R. on North American shelf fisheries. In subtropical regions, upwelling areas which may form the natural bases for stock units cross national boundaries, or lie within them. In temperate waters, the natural unit is probably a gyre like the Alaska gyral, or the Norwegian Sea. They bear little relation to national boundaries and coastal states would have to co-operate in the study of the fish stocks that enter and leave their waters. The commission structure is a useful basis for such co-operation.

There are now between ten and twenty active commissions whose business is to conserve the fish stocks throughout the world ocean. Their structures differ as do their means of taking scientific advice. There is no ideal form and all appear to work when the scientific advice is clear and well presented. In NEAFC advice comes from the International Council, in ICNAF from its Standing Committee on Research and Statistics, in the Pacific species commissions from the international directors and in INPFC from the open sessions. In all forms of institution success or failure in management can often be linked to understanding or misunderstanding in science. The failure to understand persists in scientists: Max Planck said that the only way in which a thesis was replaced was by the death of its originator. However science changes more quickly than scientists and perhaps they should spend comparatively short periods at a time on commission work.

It is sometimes suggested that because science is used for political ends, it changes its nature. If science can survive the ends it is put to in the academic field it is well adapted to do so elsewhere. But, less flippantly, nations are represented in the Commissions in order to agree, otherwise they would not meet; no purpose is served by obstruction that is not better served by absence. What cannot be done for political reasons may be stated and what can be done is discussed. The science is used to suggest action or to assess its worth. So long as knowledge is openly verifiable, it is hard to pervert. Wherever knowledge is used, it is used to some purpose. There is no difference between the use of knowledge to cure people, kill people, or merely to add to knowledge, so far as knowledge itself is concerned. The judgements made on the use of knowledge are of a different quality and do not affect the argument. Any use to which science is put must be political and all scientists should be aware of it.

6

Science and conservation

1. The failures in conservation

Fishermen live by the regularity in the season of the fisheries they exploit and by the same token they are familiar with stocks that disappear and return in long periods of time. The Norwegian and Swedish herring fisheries have oscillated for centuries and the Plymouth herring was shown to fail through natural causes. But there has been a tendency to attribute collapse to natural causes in a general sense, rather than to the effects of fishing. In the Japanese sardine fishery, three successive year classes failed when the exploitation rate was high, and the recruitment failure was attributed to a warming and diversion of the Kuroshio current. This change reversed in the early forties, but the sardine stock did not recover and recruitment continued to decline and the catches fell to very low levels indeed. Because of the nature of the stock and recruitment curve in clupeids, recruitment may have failed because the stock had been fished too hard. The year classes in the Californian sardine fishery also failed when fishing mortality was high; at the time, the decline in recruitment was correlated with failing sea temperatures an explanation that does not exclued the effect of fishing. Neither the Japanese sardine fishery nor the Californian one was subject to any commission control or regulation by the coastal state. We cannot decide exactly whether recruitment failed through natural or man-made causes because the stocks were fished at a high enough level for the two effects to be confused. Had they been exploited at the level of best recruitment, the effects might have been distinguishable and conservation would have been applied as well as possible.

The dramatic failures in conservation occurred through recruitment overfishing. Because of the nature of the stock and recruitment relationship in different groups of fishes, clupeids are more vulnerable to it than gadoids and for the same reason they are more sensitive to long term climatic changes. Both W. F. Thompson and Michael Graham believed that recruitment was independent of stock within the fishable levels of stock and for the halibut and the plaice the assumption may be very nearly true. One distinction between the Pacific Commissions and the Atlantic ones is that the Pacific stocks have not suffered disastrously from recruitment overfishing. Perhaps some of the salmon stocks were reduced for this reason in the past and the British Columbian herring stock may well have suffered recently. The deeper reason for lack of disaster in the Pacific was that the scientific model used was usually the logistic curve or some variant and its method of application was to take into account the small annual changes in stock.

The failures in the Atlantic are the collapses in the herring stocks and the dangerous state of the arctonorwegian cod stock (Garrod 1967). It is of course possible that some of the herring stocks failed naturally, as for example, the Firth of Forth spring spawners, but most failed when the exploitation rate was high enough to reduce recruitment. In the arctonorwegian cod stock, after a period of high effort, four successive year classes were very low indeed. Whereas the variability of recruitment at any one level of stock is determined by the environment, the question should not be left there. If stock is low, the managers should restrain fishing till the stock recovers, or not; if it recovers it should be exploited at a point in stock near that at which the highest recruitment per unit stock occurs. Then the greatest chance is given of distinguishing the effects of fishing from natural causes. With hindsight, it is clear that this is how the herring stocks should have been managed as they collapsed one after the other.

The collapse of the Downs stock of herring in the North Sea can be partly attributed to the inability of the scientists to grasp the implications of the stock and recruitment problem. There was, however, an added complexity. There were three stocks of autumn spawning herring in the North Sea that segregated independently to their spawning grounds and mixed on their feeding grounds. Hence it was difficult to separate fishing effort by stocks; an attempt was made that did not carry conviction but enough separation was achieved to show a stock and recruitment relationship for the Downs herring. Later the stock question was shelved and a catch quota for the North Sea was introduced which effectively assumes that all three stocks are of the same size. Conservation came too late because the scientists failed to agree on the causes of collapse: the argument was based on an assumed dichotomy between natural and man-made causes. As we have already seen such a dichotomy need not exist.

The decline of the atlantic-scandian stock raises a different question. The exploitation rate was probably high enough to reduce recruitment and the stock was in danger of recruitment overfishing. Towards the end of the last Norwegian herring period, during the last century the fishery shifted northwards from Utsira to Stadt and from Stadt to the Lofoten Islands and then the fishery collapsed and the Swedish herring fishing started. This sequence was repeated during the late 1950s and the 1960s and the fishery collapsed in 1968. From the periodicity, collapse might have been attributed to fishing or to natural causes, which implies that the wrong question was asked. The separation between fishing and natural causes is uninteresting, because the stock should be managed at the level of maximum recruitment per unit stock, so that any natural failure can be established without argument (Fig. 9).

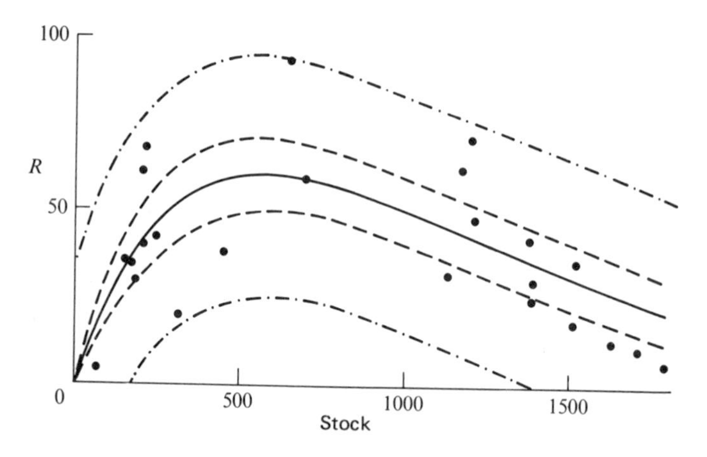

Fig. 9. The stock and recruitment curve of the arctonorwegian cod stock. The inner error lines represent those to the curve, i.e. there is only one chance in twenty that the curve can lie outside them; the outer error lines represent the errors on a single recruitment. Where the outer error lines cut the abscissa there is more than one chance in twenty of zero recruitment and at such positions the effects of fishing and of natural causes can become confused. At the highest recruitment, any failure due to natural causes should be immediately distinguishable from the effect of fishing in that the recruitment should fall below the outside error line under natural causes but not by fishing. Any prudent manager would seek to maintain the stock at the maximum recruitment per unit stock.

The reason for the Atlantic failure originated in the small plaice problem and the solution to the problem of growth overfishing. The yield per recruit model that provided the solution assumed that recruitment would not decline within the fishable levels of stock. The high variability of recruitment was taken to justify this assumption, but it also concealed the point in stock at which recruitment declined and by the time the year classes were seen to fail, it was already too late. To further complicate the issue, at such low levels of stock the high variability of recruitment often makes it impossible to distinguish the effects of fishing from environmental causes. The high variability led the scientists into the dangerous course of assuming that recruitment was independent of parent stock, whereas a more prudent procedure would have assumed some relation between them.

The fall of the blue whale stock and its potential resuscitation is the classical example of recruitment overfishing. The problem was grasped at an early stage in the decline but the solution failed to carry conviction. Action was taken when the solution was put in a clear and simple manner. Similarly, it was the simplicity of the yield-per-recruit model that carried conviction in the Atlantic commissions and led to the strong institution that exists today. The Atlantic failures in conservation were due to the inability of scientists to understand the nature of recruitment variation rather than to the shortcomings of the yield-per-recruit model. In the Pacific no such failures occurred within the commissions because the descriptive model used took into account of the changes of recruitment with stock. The Pacific failures are all outside the commission structure which is institutionally rather weak. History took different courses in the two oceans, and neither was right or wrong. In the future the Pacific commissions might embrace all the resources and the Atlantic ones should prevent recruitment failure.

The failures in conservation indicate the link between science and management. The first steps in conservation were taken because people like W. F. Thompson and Michael Graham were able to express the descriptive model of the logistic curve in the clear and simple way that carried conviction. For the same reason the yield-per-recruit model carried all before it in the early stages of the Atlantic commissions. The failures in conservation, the blue whales, the North-east Atlantic herring and perhaps the arctonorwegian cod, were not institutional failures, but were due to the inability of the scientists to keep up with events. There should be no dismay that scientists admit these failures publicly. In their private and unpublished lives scientists, by their nature, live with failure with the occasional enlightenment of success. Hypotheses survive till disproved and their life spans vary. Management should not treat the scientists' advice as a sacrosanct certainty but as an assessment of probabilities, some right and some wrong, just as in any other sphere of life.

2. The problem of stock and recruitment

The great advantage of the yield per recruit solution to the problem of growth overfishing is that management can be given answers quickly without having to wait for the slow accumulation of yearly observations that is needed in the descriptive model. At the present stage of scientific development, reasonable descriptions of the stock and recruitment relationship can be elaborated if there are 20 or 30 years of observations. The step needs to be made to some form of analytic model, like that developed by Beverton and Holt, by which advice can be given immediately, because the present rates of discovery and of exploitation are high.

In the first chapter on the natural history of fishes the nature of the density-dependent processes was described, particularly how in growth and mortality they decline with age. In the exploitation of the food available, growth and mortality appear to be linked in that fish that fail to feed and grow are taken by the everpresent predators. The major losses in numbers occur in early life as do the more important increments of weight. In the Beverton and Holt solutions to the problems of the dynamics of fish populations, the death rate was integrated in age with the growth rate in the adult age group. A deceptively simple way of formulating the solution to the stock and recruitment problem is to extend the integration from the earliest larval age groups. The rates of growth and death in the adult age groups are easily estimated and if such generalizations could be created for the life in the larval drift and on the nursery ground, the problem would be solved quite readily, that is to formulate the changes in growth and death with age in terms of available food.

At first sight the demand is impossible and may appear to be beyond our present scientific scope merely because food availability varies from year to year under the stress of climatic change. However, as described earlier, within the terms of reference of a simulated model, average conditions can be described. In model terms, it is possible to generate the density dependent mortality needed from a stock–recruitment relationship derived from many years of annual observations. There is a very important point here in that the average condition is that needed to describe any stock–recruitment curve from which any individual year class represents a deviation. Hence there is some possibility of developing a stock–recruitment relationship independently of the annual year classes.

However, little is known of the basic biology of the larval and immature stages of many commercial species. The exact locations of spawning grounds and nursery grounds are known only sketchily. The growth rates of the larvae of a few species are known, but the death rates have been established satisfactorily for only one or two species. Similar figures are needed for a number of stocks and species to establish ratios of growth to mortality and to relate such estimates to the adult weights and fecundities. The absolute death rate must differ considerably from species to species, but within a stock the ratio of growth to mortality must vary with the quantity of food available. However such are the research problems and to some extent their outcome cannot be foreseen.

In the previous section the successes and failures of conservation were linked to the degree of scientific success. Many of the organizations concerned with conservation also co-ordinate research. It may be the detailed technology of the stretch of synthetic trawl twines, an international ground fish survey or combined experiments on gear selectivity. Such work is always needed in one form or another. However the research demanded by the international institutions is usually of the applied or technological character that yields an expected answer. The other form of research at the edges of our knowledge is carried out in national institutes and it progresses slowly from an international point of view in a disjointed and dissociated manner. This does not mean that the great problems facing scientists are not discussed internationally; the International Council held a number of meetings on the population dynamics of fishes in the 1930s and quite recently held a valuable one on the problems of stock and recruitment. FAO has a long history of arranging valuable meetings on the various aspects of fisheries science, on sardines and tuna, on acoustic methods of use in research and on food-chain studies. ICNAF held a number of useful meetings on the role of the economic studies in fisheries management. Such meetings are very valuable but represent the progress made in the national institutes and it is here that the basic problems of population dynamics are disentangled.

The successes and failures in management are not discussed in the commissions. Scientific advice is proffered and taken as the best at the time and place. Advice given to NEAFC on the collapse or the herring fisheries in the northeast Atlantic did not lead to action, and it was tacitly assumed that the causes of failure were inaccessible or beyond the capacity of the science at the time. The Tuna commission has not asked why the catches are so much greater than the estimated maximum sustainable yield. The successful programme of catch quotas in ICNAF depends upon cohort analysis and there is a need to examine the variance of catches and the sensitivity of the method itself. All these problems are the common gossip of the scientists when they meet. The link between science and management suggests that the quality of management depends on success and failure more selfconsciously than in the past and it is in the national institutes that success and failure should be assessed.

The problem of stock and recruitment is at the centre of fisheries science today and its solution needs an active scientific pursuit. The traditional way is to allow the solution to appear slowly in the national institutes until it becomes part of accepted science. However the information needed can only be obtained at sea and a co-ordination of research in different parts of the world might accelerate progress. Such co-operation does take place in an informal way and the first need is the realization that the problem needs a solution. Specifically the solution should be an analytic one resembling that of Beverton and Holt twenty years ago, so that management can take decisions quickly without the tiresome accretion of annual observations.

3. The study of recruitment

As noted so frequently the variability of recruitment is high. However there is only a limited number of stocks for which adequate series of observations are available, that is, for twenty years or more. In the Downs stock of herring and the Karluk river salmon (in Alaska) variation in recruitment is comparatively low, by a factor of three to ten. The Norwegian herring recruitment and that of the arctonorwegian cod stock varies by one order of magnitude or a little more. In the North Sea haddock and the North Sea sole the year class strength can vary by as much as two orders of magnitude. A reasonable generalization would be that the annual increment of recruits can vary by one or two orders or magnitude. The variation in stock is less, by the square root of the number of age groups. But fishing reduces the number of age groups in the stock and with increased fishing pressure the variability of stock increases. Whatever the nature of the stock the annual increment of recruits forms the most important component of stock variation and if stocks are to be regulated by catch quota some form of forecasting system is desirable, as for example the systems analysis study of the Bristol Bay salmon.

So far the variation in recruitment has been considered as varying about a mean. It would be more realistic to envisage it as a variable about a trend, sometimes upward and sometimes downward, for recruitment is responsive to climatic changes; for a period of more than 50 years, catches of the arctonorwegian cod stock in the Vestfjord were shown to be highly correlated with differences in the annual increments of growth on pine trees in the area. Such a correlation can only make sense if the cod and the pine trees respond to the same unspecified climate factors. The series of year classes in the Norwegian herring and the Karluk salmon show slow upward or downward trends in time. In 1962 in the North Sea a haddock year class appeared that was 25 times larger than any of its predecessors in a time series since 1922. The climate in north-west Europe changes in a cyclical manner and there is a sense in which it may be said that the fish stocks are able to rectify such changes, as for example in the alternation of the Norwegian and Swedish herring periods. The detailed prediction of each year classes may be impossible, but it would be useful to forecast the great changes such the presence or absence of the Norwegian herring or the dramatic ones like the upsurge of haddock, cod, and whiting in the North Sea during the sixties or just the large year classes and the poor ones. However the same mechanisms are required to forecast the great changes as for the details.

The earliest form of forecasting appeared in the market sampling system itself. In many fish stocks the recruits enter the fishery in successive ages, the bigger recruits first in the youngest age group. An abundant year class is first noticed as being numerous in that age group and a scarce one may in fact be absent from it. In this way a rough method of forecasting can be developed which provides one or two years' warning of good or bad year classes. A later method was to estimate the numbers of fish in their first or second year on the nursery grounds. The major part of the density-dependent processes is by then complete and good forecasts of the magnitude of recruitment can be obtained; the disadvantage of the method is that the work has to be done at sea and if applied to a large number of fish stocks would be very expensive in research vessel time. However it has been developed successfully on the Barents Sea nursery grounds of the arctonorwegian cod stock; a successful prediction was made of the recent four very low year classes that have so reduced the stock. It could be applied to inshore nursery grounds such as that of the plaice on the Waddensee in northern Holland which are sampled with pushnets. However not all nursery grounds can be visited so easily and a considerable research vessel budget would be required for the remainder.

The question arises whether recruitment can be forecast without going to sea. In the first chapter it was pointed out that in temperate waters fishes appeared to spawn at a fixed season to take best advantage of the production cycle for their larvae. The production cycle varies in timing from year to year and the quantity of food available to the larvae may depend on the match or mismatch of larval production to that of their food. The production cycle, or the spring outburst, of temperate waters has been described in many parts of the world ocean, but usually in the more convenient places. Rarely has it been described more than once in one position, although indices of its variability have been derived from the plankton recorder network, which consists of monthly samples along the fixed tracks of certain merchant vessels.

However, in recent years knowledge of the production mechanism has increased considerably. It is essentially a predator–prey system in which herbivorous copepods feed on planktonic algae. Production starts in early spring as the sun's altitude increases and as the wind slackens. The reproductive rate of the algae increases and at a fixed food level the herbivores spawn. The peak of the cycle is reached when the reproductive rate of the algae equals the grazing rate of the herbivores. The full cycle is described by a bell-shaped curve in time and the environmental determinants such as the sun's radiation that is modulated by cloud cover and the depth of mixing that is affected by the strength of the wind. Such information is available in large quantities and the production cycle may be simulated fairly easily provided certain simplifications are made, for example, to express the algal numbers as quantities of carbon.

It is already known that recruitment can be correlated with climatic factors and the only way in which such correlations can be explained is in the modulation of the production cycle as suggested in the last paragraph. In principle the models are easy to construct and they have been in use in one form or another since 1939. However to make them accurate requires more information on the biology of the algae and the copepods that drive the cycle. Traditionally this sort of research has been outside the province of fisheries biologists, being practised in the main by marine ecologists and marine chemists. Just as population dynamics has switched from the adult life history to the juvenile stages in the search for a solution to the problem of stock and recruitment, so in the problem of recruitment variability we must look at the generation of the larval food.

The problem of year-class variation is very close to that of the dependence of recruitment on parent stock, so close that they are probably mediated by the same single process, the exploitation of food by the larvae and the young fish. If commissions are to move to catch quotas as their main regulatory weapon, research is needed on the lines indicated above. However there is a difference between the two cases, the stock and recruitment problem and the variability of recruitment. In the first case, commissions might be interested in a general way to stimulate the science within the national institutes; but in the second case, the use of catch quotas demands accurate information and one might expect commissions to be more actively concerned. The method in the study of the Bristol Bay salmon has already been referred to and in Britain recruitment forecasts are already in use as the information emerges from the market sampling and as estimates appear from the surveys of young cod in the Barents Sea. To summarize, there is a general need to be able to indicate the large scale climate changes that have modified the fortunes of the fisheries to a considerable extent and at the same time forecasts of some fluctuations are required on an annual basis to accommodate the use of catch quotas within the commissions.

4. Statistics

The bones of fisheries research are the catch statistics, the market measurements and, where possible, the age determinations. The methods of collecting statistics vary considerably. For example, in Britain the mate of a fishing vessel is asked how much he has caught and where he has been and how many hauls he has made. The system is based on the assumption that the vessel visited a single ground and then its average catch represents a good estimate of stock density on that ground. In the early years of this century this assumption was justifiable, but even within the North Sea boats are liable to visit more than one ground in a trip depending upon the market information they receive on the radio. The ideal system is to use a log book in which the quantity of each catch by species is entered by position; the log book should be so designed as to facilitate access by a computer. The Japanese have logged their tuna catches by 1° squares in all the world's subtropical and tropical oceans and the distributions of catches by species and seasons show the movements of the fisheries in elaborate detail. In recent years, the accuracy of some Japanese estimates of the vital parameters of fish populations from the Far Seas Research Laboratory at Nankai (where the tuna are studied) has been high and it may be attributed to the high quality of data collection.

It is sometimes suggested that skippers should not be asked to collect information in such detail because the secrets of their skills are being given away. All the material should be processed automatically and whereas it might be possible to rob the machines of information, the skippers should be assured that it is safeguarded. The real safeguard should be that the information is published in the form of an age distribution which must be quite innocuous from any competitive point of view.

At the present time statistics are compiled rather slowly and it is not an uncommon experience for fisheries biologists to be working a year or so behind. This does not matter if stocks are conserved by closed season or by mesh regulation. Where a catch quota is used on a single stock, as in the Fur Seal Commission or the Tuna Commission, the system is small enough to work slowly, even if both use automatic data processing today. Let us imagine that each stock in each area of the Atlantic has a separate catch quota and all are exploited by a mobile fleet. Each quota is reached at a different point in time and skippers would compete in not getting caught by reaching it.

The only way eventually to overcome such difficulties is to run the system automatically and rapidly. At the present time, catch quotas are split between countries and there is no need to process the data internationally, particularly as the quota is imposed separately on the national fleets. But there might be a considerable case for interchanging the biological statistics. The present system in the Atlantic is for scientists to take information to an international working group where it is analysed in concert. The disadvantage to this system is that it is time-limited and time is spent on merely comparing catch data. Consequently the scientists tend to take the common agreed answer that may be the least solution. With common computers and common programmes, with a continuous data input and a telex network, there is no reason why such a working group should not continue for much longer periods in separation: scientists might then meet for two days to write their reports.

Eventually the forecast catches will have confidence limits so that catch quotas can be set as accurately as possible. As cohort analysis is well adapted to estimating catches in absolute numbers in stocks with long age distributions, the variability inherent in such an analysis needs to be estimated. This is not so much the errors or biases in method that have already been examined in a preliminary way, but the consequential analysis of error arising in the catches themselves. If they are split into small units of area, species, age, and time, the overall variance of the system can be estimated. The efficiency of conservation by catch quota is then a function of the cost of data collection and data processing. National industries may demand as accurate catch quotas as possible whereas an international commission might be content with a less accurate one that achieves the end of conservation equally well.

The data handled in the international commissions are sound but often they are not fully exploited. Their quality could be improved considerably by the use of log books in which are recorded details of each catch by position of capture. Then, rather than using lumped or averaged material, the data can be used in full to the limits of the distributions. Ideally they should be handled on common computer systems as quickly as possible. Progress towards such an ideal may be a little slower than might be imagined, primarily because the scientists can obtain reasonable answers by overwork. Much better solutions could be obtained more easily with automatic data processing.

5. The future of management

The first steps in management were the prevention of disaster to secure the resources for further exploitation. By 1930, the Pacific halibut fishery must have become so unprofitable that the vessels were only kept at sea by the world-wide slump. By 1973, the stock of blue whales in the antarctic had probably been saved from extinction with the prospect of exploitation again in the decades to come even if such catches are still thirty to fifty years away. Later in the field of conservation, managers hoped to obtain optimal yields, maximum sustainable yields, or the best economic yields. Maximum catches have been obtained, but optimal economic yields probably still elude our best efforts. Since 1930 a number of collapses have not been prevented, chiefly of clupeid fishes, and the series culminated in the recruitment failure in the Peruvian anchoveta stock in the last two years. To some degree, fisheries science has promised more than it could produce. The failure to match achievement with promise is a common one and if we are surprised that scientists also fail, it is because they are sometimes regarded idealistically and not realistically. In 1971 Gulland noted that fisheries administrators had come to expect conclusions with a stamp of certainty. However this state arose, it is an unhealthy one, being much too demanding on the scientists and too sheltering for the administrators. They are trained to judge probabilities and possibilities and can do so perfectly well on conclusions based on scientific evidence.

The most important point established is that science and management are indissolubly linked. Failures in conservation were attributed to scientific failures and success in management to scientific success. The judgement of success in the science is not the same as the judgement of good or bad science, although they may sometimes coincide. For example the analytic models of Beverton and Holt are scientifically of a higher level than the descriptive models of an earlier generation, merely because more information was subsumed in a single expression. Initially the yield-per-recruit model was successful because it solved the problem of growth overfishing which had resided in the minds of European scientists since the time of Petersen. More important the solution carried conviction to managers particularly when associated with mesh regulation. The subsequent failure in those stocks that suffered from recruitment overfishing lay not in the scientific quality of the Beverton and Holt models, but in the enthusiastic application of the yield-per-recruit model, a little blindly, rather than the use of the self-regenerating yield curve, or some analogue. Had such procedures been developed, the phrase recruitment overfishing might never have been used. Thus the criterion of scientific success in the fisheries field is really one of management, that is, whether it failed or succeeded and it is quite distinct from any criterion of scientific quality. Not only are the successes and failures of management linked to good and bad science, but scientific success is assessed in terms of management.

It is often said that a particular course was not taken for political reasons, as if it were improper for a country to protect its own interests in the face of an ideal scientific solution to a problem. For example, the Japanese pursued the blue whale to a dangerously low level perhaps because the British, the Norwegians, and the Dutch had unloaded their whaling fleets on to them in a falling market. It must always be remembered that the science can only be of use within the fields on which the participating countries are prepared to agree. If there is disagreement on grounds that are purely political it is unlikely that the science can be used to obtain agreement. However there are cases when a country might take a stance because the science has been improperly understood. The failure to manage the north-east Atlantic herring stocks appeared to have political components. If the scientific case against recruitment overfishing had been put clearly and convincingly with the means of preventing it in 1956, the political difficulties might have disappeared. No country can think of a good political reason for scrapping a profitable industry worth millions of pounds. However, any fisheries scientist returning from an international meeting is only too aware of the political overtones that have reverberated around him. But such overtones are common to any application of scientific knowledge.

In the future, management may be expected to spread rapidly across the remainder of the world oceans and the number of international commissions may be augmented. They are successful enough to have formed the very advanced international structure that exists in the North Atlantic yet strong enough to absorb the scientific failures associated with recruitment overfishing. In other parts of the world there is every reason to expect that the same sort of resilient structure will emerge, in differing forms in different areas. At the present time, before the coming Law of the Sea Conference, there is much discussion on the rights and needs of coastal states. It is not a question that concerns science or management because both must continue. All that can be said is that the stocks must be monitored and as most must be shared by states including coastal states, an international structure of management would still be needed. In an ideal world the present system of international commissions would survive.

It is sometimes suggested that the population dynamics of fishes is a science that has been completed. A science is never finished, but some last longer than others: Newtonian mechanics will be of use long after its scientific bases have been supplanted. One of the disadvantages of the descriptive models of populations is that the control mechanisms are accepted but ignored, whereas they should be understood. But the present demands of management in using catch quotas require that the control mechanisms and the fluctuations of year classes are fully understood and this is a research problem of considerable magnitude. In the long term as the fish stocks respond to or insulate themselves from the effects of climatic changes, the mechanisms of response will have to be unravelled. These, however, are problems visible to a biologist who happens to look in this direction.

A much more important point is this. If the quality of management depends on the nature of the science, so does the development of the science depend upon the demands of mangement and the long term needs of the industry and the fish eating population. Already the system analysts are used in the management of the Bristol Bay salmon fishery. In the future the main function of such procedures may be to change the job of management from many complicated choices to a few simpler ones. One might expect economic considerations to play a much more important part once an accurate system of catch quotas has been established. Whatever the future mechanisms of control will be, the demands on the accuracy and consistency of the population dynamics of fishes will increase and all the statistical needs outlined in the previous section will be fulfilled.

As a science fisheries biology is a little weak, but it is growing stronger. The success of management so far is as much an indication of man's desire to make conservation work as it is due to the science itself. In a highly-developed science an important and essential test of discovery is whether and how it fits into the existing jigsaw of knowledge. In fisheries science such a structure is a little tenuous although it is developing. For example, our best information on the death of fishes in the absence of fishing is not much more valuable than a series of guesses. Many fisheries biologists are pleased with the progress of their science and judged by the overall success of management, as a result of their science, they have some right to be. The science will, however, mature only with a little divine discontent because the discontent is the secret of divination.

Selected bibliography

Anon (1964). International Commission on Whaling, 1964. Fourteenth Report of the Commission (covering the fourteenth fiscal year 1 June 1962 to 31 May 1963). *Rep. Int. Comm. Whal.*

Beverton, R. J. H., and Holt, S. J. (1957). On the dynamics of exploited fish populations. *Fish. Invest. Lond.* 2, 19.

Cushing, D. H. (1968). *Fisheries Biology.* University Wisconsin Press, Wisconsin.

– (1972). *A history of the International Fisheries Commissions. Proc. R. Soc. Edinburgh* (*B*), 73(**36**), 361–90.

Garrod, D. J. (1967). *Population dynamics of the arcto-norwegian cod. J. Fish. Res. Bd. Can.* **24**, 145–90.

Graham, G. M. (1935). Modern theory of exploiting a fishery, and application to North Sea trawling. *J. Cons. int. Explor. Mer.* **10**, 264–74.

Graham, M. (1943). *The fish gate.* Faber, London.

Gulland, J. A. (1972). *The fish resources of the ocean.* Fishing News (Books) Ltd.

Harden Jones, F. R. (1968). *Fish migration,* Arnold, London.

Lack, D. (1954). *The natural regulation of animal numbers.* Clarendon Press, Oxford.

Ricker, W. E. (1958). Handbook of computations for biological statistics of fish populations. *Bull. Fish. Res. Bd. Can.* 119.

Schaefer, M. B. (1957). A study of the dynamics of the fishery for yellowfin tuna in the eastern tropical Pacific Ocean. *Bull. Inter. Am. Trop. Tuna Commn.* 2, 245–85.

Thompson, W. B. (1950) The effect of fishing on stocks of halibut in the Pacific. *Publs. Fish. Res. Inst. Seattle.*

Index